SOFTWARE PROJECT MANAGEMENT

Er. RISHABH ANAND
HCL TECHNOLOGIES LTD.
ITILV3 & PRINCE2 CERTIFIED
PROFESSIONAL

S.K. KATARIA & SONS®
Educational Publishers
4885/109, Prakash Mahal, Dr. Subhash Bhargav Lane,
Opposite Delhi Medical Association, Daryaganj, New Delhi-110002 (INDIA)
Phone: +91-11-43551243, Mobile +91-9871775858, 7678425209
e-mail: katariabook@gmail.com, katariabooks@yahoo.com
Website: www.skkatariaandsons.com
(e-books, subscriptions and bulk buying also available on our website)

Published By:

S.K. Kataria & Sons®

4885/109, Prakash Mahal, Dr. Subhash Bhargav Lane,
Opposite Delhi Medical Association, Daryaganj, New Delhi-110002 (INDIA)
Phone: +91-11-43551243, *Mobile:* +91-9871775858, 7678425209
e-mail: katariabook@gmail.com; katariabooks@yahoo.com
Website: www.skkatariaandsons.com

First Edition: 2012, Third Edition: 2014; Reprint: 2026

Laser Type Setting : Sara Assignments, Delhi

Printed at : Nisha Enterprises, Sahibabad

PREFACE TO THE THIRD EDITION

Revising any book is a daunting task, since authors have to consider the feedback they received from teachers and students and incorporate the same in the revised edition of the book.

In this edition, the author has made sincere efforts to make the book up-to-date. The present edition has been thoroughly revised and a lot of useful material has been added. The mistakes and misprints, which had crept in last edition, have been eliminated in this edition. The author hopes that all these features will make the book even more useful and attractive to the readers.

Last but not the least, I would like to mention that no one is 100% accurate in this world, so any suggestions for the further improvement of this book will be highly acknowledged.

Er. Rishabh Anand

PREFACE TO THE FIRST EDITION

One of the most important elements of a software development project is requirements management (RM)—a systematic approach to eliciting, organizing, documenting, and tracking a system's requirements. Proper RM helps validate and verify the system, manage change, and analyze the project's status. It's much less expensive to fix a problem during requirements analysis than during design, testing, or production. Regardless of this fact, RM is often neglected in projects, and time spent on it is proportionally too small.

Since the earliest days of the computer industry, experience has shown that managing a software projects is a complex and demanding activity. Repeated studies have shown that management problems remain one of the top two or three causes of difficulties in, or even outright failure of, software development projects. This fact has remained unchanged as the technology used to design, implement, and deploy software systems has undergone rapid and widespread evolution. Likewise, the importance of software project management has been unaffected by changes in the business and structural models employed by both industries and individual companies. Thus, while the fundamental issues of software project management have consistently exerted an influence on the success of product developments, the same mistakes continue to be made:

- Requirements are unclear at the beginning of projects and are not managed during the project.
- Project planning is incomplete or not done at all.
- Management of the project is weak or arbitrary.
- Underlying and supporting processes needed to perform the technical work are often ignored.

It seems individual companies and software managers have had to re-learn these same lessons. Yet, the profession as a whole has recognized that certain issues are of fundamental importance and has made various attempts to capture the 'best practices' that can be used to address these problems.

The book follows a 'tall and thin' approach. The book provides logical method of explaining various complicated concepts and stepwise methods to explain the important topics. Each chapter is well supported with necessary illustrations and practical examples. All care has been taken to make students comfortable in understanding the basic concepts of the subject.

The book not only covers the entire scope of the subject but explains the philosophy of the subject. This makes the understanding of this subject clearer and makes it more interesting. The book will be very useful not only to the students but also to the subject teachers.

The author hopes that the book will fulfill the pressing need of interested readers and welcome any suggestions towards the improvement of the book.

Er. Rishabh Anand

July, 2012

ACKNOWLEDGEMENT

Books are not simply cast into form. Besides the author, there are usually many more people who contribute in terms of ideas and discussions, and in the form of technical as well as emotional support. I cannot list all who helped to sustain me over the course of writing this book, and I apologize for any possible omissions.

We all marvel at the beautiful rose. The rose, with all its beauty and grandeur. But seldom we pause and thank the kind of gardener who patiently manured and watered it. The book I have written is the fruit of such unseen hands.

I am especially grateful to my teachers and my colleagues specially Mrs. Aparna Mahajan, Mr. Sanjeev Kumar, Mrs. Rashmi Mahajan, Wg.cdr. P. Patnayak, GpCapt. H.S. Dua, Wg.cdr. Onkar Singh, Mrs. Dimple Saproo, Mrs. Naresh Kumari, Mrs. Chitra Kaul, Director M.M Jha, Mrs. Savita Khazanchi, Mrs. Meenakshi and other faculty members who influenced my academic grown during my study.

There was, in and always will be the love of my students whom I taught, for what is written in this book has been the fruit of experience that I taught them. Finally, thanks to my mother and to rest of my family for their patience and support during the long hours of writing this book and above all there is the one almighty whose humble children we are. It is his blessings we cherish and pray for. It is the blessing I wish for you.

I express my deep thanks to Mr. Sanjeev Kataria, the Managing Director of M/s S.K. Kataria & Sons as well as his entire team for publishing the book in such a beautiful get-up and well in time.

When you finish the journey through this book, having variety of experiences, ideas and thoughts. You are welcome to express your opinion at **katariabooks@yahoo.com**

Er. Rishabh Anand

Dedicated to

My Mother whose love and affection made me capable of writing this book

And

Shri Guru Maharaj Ji of Shri Anandpur

CONTENTS

CHAPTER 1

INTRODUCTION TO SOFTWARE PROJECT MANAGEMENT

INSIDE THIS CHAPTER

1.1 INTRODUCTION

Software Managers are responsible for planning and scheduling software project development. They supervise the work to ensure that it is carried out to the required standards.

This textbook is about 'Software Project Management'. The first question is whether the management of software projects is really that different from that of other projects.

In the mid-1990s, at least three important analyses of the state of the software engineering industry were performed. The results were presented in Patterns of Software Systems Failure and Success [Jones, 1996] in "Chaos" [Standish Group, 1995] and in Report of the Defense Science Board Task Force on Acquiring Defense Software Commercially [Defense Science Board, 1994].

All three analyses reached the same general conclusion: The success rate for software projects is very low. Although the anlayses had some different perspectives, their primary message were complementary and consistent.

Software project management is the art and science of planning and leading of software projects. It is a sub-discipline of project management in which software projects are planned, monitored and controlled. What exactly do we mean by 'software project management'? To answer this, we need to look at some key ideas about the planning, monitoring and control of software projects. Projects to produce software are worthwhile only if they satisfy real needs and so we will examine how we can identify the stakeholders in a project and their objectives. Identifying those objectives and checking that they are met is the basis of a successful project. This, however, cannot be done unless there is accurate information and how this is provided will be explored.

1.2 FUNDAMENTALS OF SOFTWARE PROJECT MANAGEMENT (SPM)

(*MDU 2008-09, 2011-12, 2012-13*) (*10 Marks*)

A software project has two main activity dimensions: engineering and project management. The engineering dimension deals with building the system and focuses on issues such as how to design, test, code and so on. The project management dimension deals with properly planning and controlling the engineering activities to meet project goal for cost, schedule, and quality.

Software project management is an umbrella activity within software engineering. It begins before any technical activity is initiated and continues throughout the definition, development, and support of computer software. Software project management encompasses the knowledge, techniques, and tools necessary to manage the development of software products.

The goal of Project Management is to ensure that a software product that meets the quality specifications of the requirements definition is produced on schedule and at economically justifiable cost within the planned resources (e.g. personal, equipment, tools, etc.).

1.2.1 Software

Software is described by its capabilities. The capabilities relate to the functions it executes, the features it provides and the facilities it offers. Software written for sales-order processing would have different functions to process different types of sales orders from different market segments. The features, for example, would be to handle multi-currency computing, updating of product, sales and tax status in MIS reports and books of accounts.

The software is developed keeping in mind certain hardware and operating system considerations, known as platform. Hence, software is described along with its capabilities and the platform specifications that are required to run it.

The best thing about software is its flexibility: It can be programmed to do almost anything. The worst thing about software is also its flexibility: The "almost anything" characteristic has made it difficult to plan, monitor, and control software development. This unpredictability is the basis of what has been referred to for the past 30 years as the "Software Crisis".

Definition of software

Software is a set of instructions to acquire inputs and to manipulate them to produce the desired output in terms of functions and performance as determined by the user of the software. It also includes a set of documents, such as the software manual, meant for users to understand the software system. Today's software comprises the Source Code, Executables, Design Documents, Operations and System Manuals and Installation and Implementation Manuals.

Software is:

(*i*) Instructions (computer programs) that when executed provide desired function and performance.

(*ii*) Data structures that enable the programs to adequately manipulate information.

(*iii*) Documents that describe the operation and use of the programs.

OR

The term software refers to the set of computer programs, procedures, and associated documents (flow charts, manuals, etc.) that describe the programs and how they are to be used. To be precise, software means a collection of programs whose objective is to enhance the capabilities of the hardware.

OR

Definition of software given by IEEE

"Software is the collection of computer programs, procedure rules and associated documentation and data."

Importance of Software

(Important)

Computer software has become a driving force.

- It is engine that drives business decision-making.
- It serves as the basis for modern scientific investigation and engineering problem solving.
- It is embedded in all kinds of systems like transportation, medical, telecommunications, military, industrial processes, entertainment, office products etc.

It is important as it affects nearly every aspect of our lives and has become pervasive in our commerce, our culture and our everyday activities software impact on our society and culture is significant. As software importance grows, the software community continually attempts to develop technologies that will make it easier, faster and less expensive to build high-quality computer programs.

1.2.2 Project

(*MDU 2009-10, 2012-13, 2013-14*) (*10 Marks*)

The dictionary definitions put a clear emphasis on the project being a planned activity.

The word "project" was first used in or around the sixteenth century and derives from the Latin projicere (= throw forward). The Latin root thus suggests movement, a trajectory, a certain relationship with space and time. A project is a unique venture with specific start and end dates. This is different from an ongoing task that doesn't have an end date. Projects are run by people and often involve different parts of an organization. Constraints on project include cost, schedule, resources, and quality. There is a give and taken between these items *i.e.* you can't have it all. Usually projects are divisible into stages or phases each with their own set of priorities and goals.

The term project has a wider meaning. A project is accomplished by performing a set of activities.

Definitions

A project is a temporary activity characterized by having a start date, specific objectives and constraints, established responsibilities, a budget and schedule, and a completion date.

OR

A project is a set of related tasks that are coordinated to achieve specific objectives in a given time limit.

OR

"*A project is a series of activities or tasks that have a specific objective to be completed within certain specifications, have defined start and end dates, have funding limits (if applicable), and consume resources (i.e., money, people, equipment).*"

—Harold Kerner

OR

James Lewis views a project "*as a one-time job that has definite starting and end points, clearly defined objectives, scope and a budget ; differentiated from repetitive activities such as production, order processing and so on ; a special activity with very tactical goals.*"

OR

The Oxford English Dictionary defines project as "*An individual or collaborative enterprise that is carefully planned and designed to achieve a particular aim ; [e.g.,] a research project/a nationwide project to encourage business development.*"

OR

A temporary endeavour undertaken to create a unique product, service, or result.

OR

A project, by definition, is a temporary activity with a starting date, specific goals and conditions, defined responsibilities, a budget, a planning, a fixed end date and multiple parties involved.

Characteristics of a Project

(Important)

The various characteristics of a project are as follows:

1. **Projects have a purpose.** Projects have clearly-defined aims and set out to produce clearly-defined results. Their purpose is to solve a "problem", and this involves analyzing needs beforehand. Suggesting one or more solutions, it aims at lasting social change.

2. **Projects are realistic.** Their aims must be achievable, and this means taking account both of requirements and of the financial and human resources available.

3. **Projects are limited in time and space.** They have a beginning and an end, and are implemented in a specific place and context.

4. **Projects are complex.** Projects call on various planning and implementation skills, and involve various partners and players.

5. **Projects are collective.** Projects are the product of collective endeavour. They are run by teams, involve various partners and cater for the needs of others.

6. **Projects are unique.** All projects stem from new ideas. They provide a specific response to a need (problem) in a specific context. They are innovative.

7. **Projects are an adventure.** Every project is different and ground-breaking, they always involve some uncertainty and rise.

8. **Projects are made up of stages.** Projects have distinct, identifiable stages.

9. **Projects can be assessed.** Projects are planned and broken down into measurable aims, which must be open to evaluation.

Prerequisites of a Project

According to an American thinker **John Dewey,** a project has four prerequisites:

- a communal reflection process, which shapes its growth and development;
- observation of the conditions in the environment where it is devised;
- knowledge of what has happened in similar situations in the past;
- an approach which synthesizes observation of the present and knowledge of the past and so identifies their meaning.

Thus, a project:

- is a method which enables us to move from idea to action, structuring the various stages in that process.
- sets out to alter the environment in which it is to take place.
- has an educational dimension and enables people to learn by experiment.
- is the product of collective activity.

Examples of Notable Projects

- Human Genome Project which mapped the human genome
- Manhattan Project, which developed the first nuclear weapon
- Polaris missile project: an ICBM control-system
- Project Apollo, which landed humans on the moon
- Soviet atomic bomb project
- Soviet manned lunar projects and programs.

Problems with Software Projects

A survey of managers is done, which identified the following commonly experienced problems during a project the following problems of software projects are from the manager's point of view:

- Poor estimates and plans;
- Lack of quality standards and measures;
- Lack of guidance about making organizational decisions;
- Lack of techniques to make progress visible;
- Poor role definition—who does what?
- Incorrect success criteria.

Below is a list of the problems identified by a number of Computing and Information System degree students who had just completed a year's industrial placement:

- Inadequate specification of work;
- Management ignorance of ICT;
- Lack of knowledge of application area;
- Lack of standards;
- Lack of up-to-date documentation;
- Preceding activities not completed on time—including late delivery of equipment;
- Lack of communication between users and technicians;
- Lack of communication leading to duplication of work;
- Lack of commitment—especially when a project is tied to one person who then moves;
- Narrow scope of technical expertise;
- Changing statutory requirements;
- Deadline pressure;
- Lack of quality control;
- Remote management;
- Lack of training.

The various problems which are faced by the customers of the product of computer projects. Some reported problematic projects have swallowed up huge amounts of money:

- The United States Internal Revenue System abandoned a tax system modernization programme after having spent $4 billion.
- The state of California spent $1 billion on its non-functional welfare database system.

Reasons why Project Fails or Succeeds

There are eight reasons why projects succeed:

1. The organizational structure is suited to the project team.
2. The project team participates in planning.
3. The project team is committed to establishing schedules.
4. The project team is committed to establishing realistic budgets.
5. The project makes proper use of network planning techniques and does not let the plan become an end at itself.
6. The project team works with bureaucracy, politics and procedures and not against them.
7. The project team agrees on specific and realistic project goals.
8. The target public is involved from the start of the project.

There are eight reasons why some projects fail:

1. Inadequate authority.
2. Lack of project team participation and planning.
3. Lack of project team participation in problem solving.
4. Inadequate communication skills.
5. Inadequate technical skills.
6. Inadequate administrative skills.
7. Unrealistic project schedules.
8. Unclear project goals.

1.2.3 Management *(Important)*

Management is defined, in simple terms as the skill or practice of controlling, directing or planning something; "The act of directing or managing for a purpose". In other words, it is the effort of planning, organizing and mobilizing people and resources for a given purpose. It involves the activities and tasks undertaken by one or more persons for the purpose of planning and controlling the activities of others in order to achieve objectives that could not be achieved by the others acting alone.

Management Activities

Management involves the following activities:

1. Planning: Deciding what is to be done.
2. Organizing: Making arrangement.
3. Staffing: Selecting the right people for the job etc.
4. Directing: Giving instructions.
5. Monitoring: Checking on progress.
6. Controlling: Taking action to remedy hold-ups.
7. Innovating: Coming up with new solutions.
8. Representing: Liasing with clients, users developers, suppliers, and other stake-holders.

Key objectives of Effective Management

A software development manager must keep in mind quality, productivity, and risk reductional throughout the planning and execution of product development.

1. **Quality:** Quality is best achieved by careful adherence to standards, effective development techniques, and periodic technical review throughout the process. Management must cooperate and coordinate with quality assurance organisations, if they exist. The Japanese emphasis on continuous quality improvement and its remarkable results offer important lessons for software developers.

2. **Productivity:** Increased productivity lower costs. In the current state of development technology, the most important productivity factors are the ability of the individual software engineers, the tools they work with, and the work environment.

3. **Risk Reduction:** Managers should identify the most difficult parts of a particular development and systematically came to grips with efficient solutions. Requirements that can become "show-stoppers" later must be dealt with early in the process.

1.3 SOFTWARE PROJECT MANAGEMENT

(MDU 2009-10, 2011-12, 2012-13, 2013-14) (20 Marks)

Software project management is the art of science of planning and leading software projects. It is sub-discipline of project management in which software projects are planned, monitored and controlled.

Definitions

Software Project Management is the process of planning, organizing, staffing, monitoring, controlling and leading a software project. **(ANSI/IEEE Std 1058.1–1987)**

OR

Software Project Management is a discipline of project is which software projects are planned, motivated and controlled.

OR

Project Management is the complete set of tasks, techniques, tools applied during project execution.

OR

Project Management is the application of knowledge, skills, tools and techniques to project activities to meet project requirements.

Objectives of SPM

The project objectives define the target status at the end of the project. The abbreviation S.M.A.R.T. is sometimes used to describe well defined objectives:

1. **Specific.** Effective objectives are concrete and well defined. Vague aspirations such as "to improve customer relations" are unsatisfactory. Objectives should be defined in such a way that it is obvious to all whether the project has been successful or not.

2. **Measurable.** Ideally there should be measures of effectiveness which tell us how successful the project has been. For example, 'to reduce customer complaints' would be more satisfactory as an objective than 'to improve customer relations'. The measure can, in some cases, be an answer to simple yes/no question, *e.g.* 'did we install the new software by 9 May?'

3. **Achievable.** It must be within the power of the individual or group to achieve the objective. Agreement with all the stakeholders what the goal should be. Is there a realistic path to achievement?

4. **Relevant.** The objective must be relevant to the true purpose of the project.

5. **Time constrained.** There should be a defined point in time by which the objective should have been achieved.

Activities of SPM

(Important)

Software project management encompasses the following activities: measurement, project estimating, risk analysis, scheduling, tracking, and control.

1. **Measurement and metrics.** To be most effective, software metrics should be collected for both the process and the product. Process oriented metrics can be collected during the process and after it has been completed. Process metrics collected during the process focus on the efficacy of quality assurance activities change management, and project management. Process metrics collected after a projected has been completed examine quality and productivity. Process measures are normalized using either lines of code or function points so that data collected from many different projects can be compared

and analyzed in a consistent manner. Product metrics measure technical characteristics of the software that provide an indication of software quality measure can be applied to models created during analysis and design activities, the source code and testing data.

2. **Project estimating.** Scheduling and budgets are often dictated by business issues. The role of estimating within the software process often serves as a "sanity check" on the predefined deadlines and budgets that have been established by management. (Ideally, the software engineering organization should be intimately involved in establishing deadlines and budgets, but this is not a perfect or fair world)

All software project estimation techniques require that the project have a bounded scope, and all rely on a high level functional decomposition of the project and an assessment of project difficulty and complexity. There are three broad classes of estimation techniques for software projects:

(*a*) ***Effort estimation techniques.*** The project manager creates a matrix in which the left hand column contains a list of major system functions derived using functional decomposition applied to project scope. The top row contains a list of major software engineering tasks derived from the common process framework. The manager (with the assistance of technical staff) estimates the effort required to accomplish each task for each function.

(*b*) ***Size-oriented estimation.*** A list of major system functions derived using functional decomposition applied to project scope. The "size" of each function is estimated using either Lines of Code (LOC) or Function Points (FP). Average productivity data (for instance, function points per person month) for similar functions or projects are used to generate an estimate of effort required for each function.

(*c*) ***Empirical models.*** Using the results of a large population of past projects, an empirical model that relates product size (in LOC or FP) to effort is developed using a statistical technique such as regression analysis. The product size for the work to be done is estimated and the empirical model is used to generate projected effort.

In addition to the above techniques, a software project manager can develop estimate by analogy. That is, by examining similar past projects and projecting effort and duration recorded for these projects to the current situation.

3. **Risk assessment, risk prioritization and risk management:** The goals of these activities are:

(*a*) to identify those risks that have a high likelihood of occurrence

(*b*) to assess the consequence (impact) of each risk should it occur, and

(*c*) to develop a plan for mitigating the risks when possible, monitoring factors that may indicate, their arrival, and developing a set of contingency plans should they occur.

4. **Scheduling.** The process definition and project management activities that have been discussed above feed the scheduling activity. The common process framework provides a work breakdown structure for scheduling. Available human resources, coupled with effort estimates and risk analysis provide the task interdependencies, parallelism, and time lines that are used in constructing a project schedule.

5. **Tracking and control:** Project tracking and control is most effective when it becomes an integral part of software engineering work. A well defined process framework should provide a set of milestones that can be used for project tracking. Control focuses on two major issues: quality and change.

To control quality, a software project team must establish effective techniques for software quality assurance, and to control change, the team should establish a software configuration management framework.

Importance of SPM

A project must successfully pass through each of the project phases in order to continue to the next. Because the organization usually commits more money as a project continues, a management review should occur after each phase to evaluate progress, likely success, and continued compatibility with organization goals. These management reviews, called phase exit or kill points, are very important for keeping project on track and determining if they should be continued, redirected, or terminated. Recall that projects are just one part of entire system of an organization. Changes in other parts of the organization might affect what is happening on projects, and what happens on projects might likewise affect what is happening in other parts of the organization. By breaking projects into phases, senior management can make sure that the projects are still compatible with needs of the rest of company.

Let's take another look at the opening case. Suppose Tom Walters' college did a study on increasing the use of technology. At the end of the concept phase, the project team could have presented information to the faculty and administration that described different options for increasing the use of technology, an analysis of what competing colleges were doing, and result of a survey of local stakeholders opinions on the subject. This presentation at the end of the concept phase represents a management review. Suppose the study reported that 90 percent of students, faculty, and staff surveyed strongly opposed the idea of requiring all students to have laptops and that many adult students said they would attend other colleges if they were required to pay for the additional technology. The college would probably decide not to pursue this idea any further. Had Tom taken a phased approach, he and his staff would not have wasted the time and money it took to develop more detailed plans.

SPM Framework *(Important)*

According to the Software Engineering Institute (SEI), a methodology must possess certain attributes in order to meet the requirements of being called a methodology another framework for methodologies is the software engineering body of knowledge that contains other knowledge development methods to be used by any professional software engineer. Software Project Management can be built on this framework as shown in Fig. 1.1.

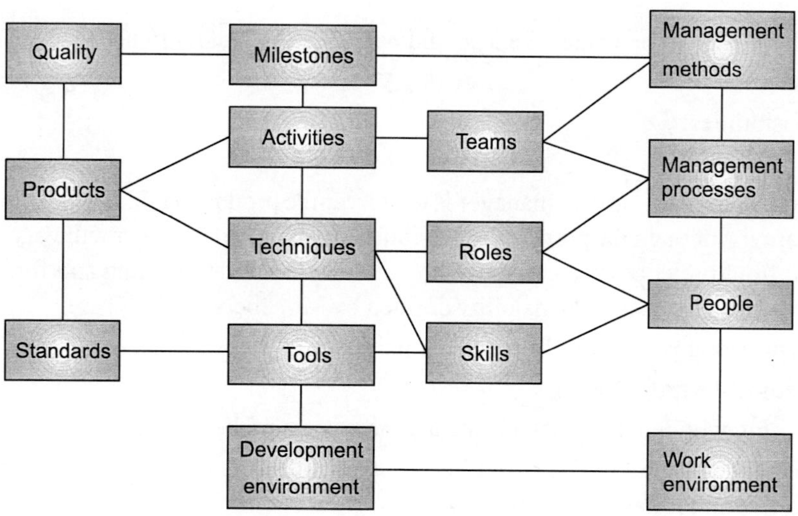

Fig. 1.1 SPM framework.

SPM framework is a process pattern view of project activities. This approach focuses on the communication and people centric aspects of project management.

1.4 CHARACTERISTICS OF PROJECT MANAGER

A project manager must be:

1. An **organiser**, with capacity to understand plan and co-ordinate efforts and resources to meet the objectives.
2. An **strategist**, able to set clear long and short-term objectives, keeping these in mind together with the reasons for the 'project's existence'.
3. A **motivator**, with skills and attitudes enabling him/her to motivate and commit people to the project or to participate in it (workers, volunteers, young people).
4. A **fund-raiser**, with knowledge and confidence to apply for funds to administer and account for them with integrity and competence.
5. An **activist**, in the meaning of someone able to spot initiatives and to organise ideas into meaningful social actions with clear values evolving over time.
6. A **visionary**, someone able to imagine social innovation and change.
7. A **community worker** with particular concern for the affairs of the community or/and organization he/she are involved in.
8. A **social worker**, capable of caring for people without replacing them, *i.e.*, instilling in them the motivation and confidence to take part in shaping their own future and realizing their projects.
9. A **teacher and a learner** capable of empowering people while at the same time being able to learn from the experiences and use that for the organization, project or community-monitoring and evaluating a process in relation to the objectives, changing plans and ultimately objectives in relation to the circumstances.

1.5 PROJECT MANAGERS ROLES & RESPONSIBLITIES (*Expected*)

When they are appointed, project managers should be given terms of reference that define their:

- objectives;
- responsibilities;
- limits of authority.

The objective of every project manager is to deliver the product on time, within budget and with the required quality. Although the precise responsiblities of a project manager will vary from company to company and from project to project, they should always include planning and forecasting. Three additional areas of management responsibility defined by Mintzberg [Ref. 17) are:

1. **Interpersonal responsibilities,** which include:
   - leading the project term;
   - liaising with initiators, senior management and suppliers;
   - being the 'figurehead', *i.e.*, setting the example to the project team and representing the project on formal occassions.

2. **Informational responsibilities,** which include:
   - monitoring the performance of staff and the implementation of the project plan;
   - disseminating information about tasks to the project team;
   - disseminating information about project status to initiators and
   - senior management;
   - acting as the spokesman for the project team.
3. **Decisional responsibilities,** which include:
   - allocating resources according to the project plan, and adjusting those allocations when circumstances dictate (*i.e.*, the project manager has responsibility for the budget);
   - regotiating with the nitiator about the optimum interpretation of contractual obligations, with the company management for resources, and with project staff about their tasks;
   - handling disturbances to the smooth progress of the project such as equipment failures and personnel problems.

1.6 SOFTWARE PROJECTS V/S OTHER TYPES OF PROJECTS

(MDU 2007-08, 2010-11, 2012-13, 2013-14) (15 Marks)

Software projects have several properties that make them very different to other kinds of engineering project:

1. **Invisibility/the Product is Intangible.** It's hard to claim a bridge is 90% complete if there is not 90% of the bridge there. It is easy to claim that a software project is 90% complete, even if there are no visible outcomes.

2. **We don't have Much Experience.** Software engineering is a new discipline, and so we simply don't have much understanding of how to engineer large scale software projects.

3. **Complexity.** Software products contain more complexity than other engineered artifacts (such as bridge or road). We can't measure the complexity of a software project until we actually work on it.

4. **Large Software Projects are Often "Bespoke".** Most large software systems are one-off, with experience gained in one project being of little help in another.

5. **The Technology Changes Very Quickly.** Most large software projects employ new technology; for many projects, this is the raison d'etre (means the purpose that justify a things existence).

6. **Working.** Software projects are based on logical work while others are based on physical work.

7. **Flexibility.** The ease with which software can be changed is usually seen as one of its strength. Software projects are flexible. Customer only wants final results, so rest of things are in control of the programmer he can modify software at any stage while this thing is not in other projects, as everything is in front of customers, he is aware about progress, he can view what work is being done by project manager and team, so it is not in the hands of project team to make changes at any stage of the project.

8. **Conformity.** The traditional engineer is usually working with physical systems and materials like cement and steel. These physical systems can have some complexity but are governed by physical laws that are consistent.

Software developers have to conform to the requirements of human clients; it is not just that individuals can be inconsistent. Organizations, because of lapses in internal communication or in effective decision-making, can exhibit remarkable 'organizational stupidity' which developers have to cater for.

9. **Stakeholders Expectations.** Stakeholders have more expectations in the software projects as compared to other types of project.

1.7 ACTIVITIES COVERED BY SOFTWARE PROJECT MANAGEMENT

(MDU 2012-13) (05 Marks)

A software project is not only concerned with the actual writing of software. In fact, where a software application is bought in 'off the shelf', there may be no software writing as such. This is still fundamentally a software project because so many of other elements are associated with this type of project are present.

Usually there are three successive processes that bring a new system into being. See Fig. 1.2.

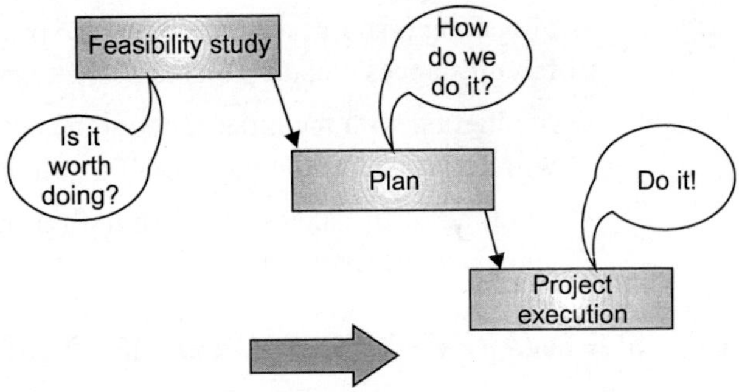

Fig. 1.2 The feasibility study/plan/execution cycle.

1. **The feasibility study.** This investigates whether a prospective project is worth starting–that it has a valid business case. Information is gathered about the requirements of the proposed application. Requirements elicitation can, at least initially, be complex and difficult. The client and other stakeholders may be aware of the problems they wish to overcome and the aims they wish to pursue, but not be sure about the means of achievement. The probable developmental and operational costs, along with the value of benefits of the new system, will also have to be estimated. With a large system, the feasibility study could be treated as a project in its own right—and have its own planning sub-phase. The study could be part of a strategic planning exercise examining and prioritizing a range of potential software developments. Sometimes an organization has a policy where a group of projects is planned as a programme of development.

2. **Planning.** If the feasibility study produces results which indicate that the prospective project appears viable, then planning of the project can take place. However, for a large project, we would not do all our detailed planning right at the beginning. We would formulate an outline plan for the whole project and a detailed one for the first stage. More detailed planning of the later stages would be done as they approached. This is because we would have more detailed and accurate information upon which to base our plans nearer to the start of the later stages.

3. **Project execution.** The project can now be executed. The execution of a project often contains design and implementation sub-phases. Students new to project planning often find it difficult to separate planning and design, and often the boundary between the two can be hazy. Essentially, design is thinking and making decisions about the precise form of the products that the project is to create. In the case of the software, this could relate to the external appearance of the software, that is, the user interface, or the internal architecture. The plan lays down the activities that have to be carried out in order to create these products. Planning and design can be confused because at the most detailed level, planning decisions are influenced by design decisions. For example, if a software product is to have five major components, then it is likely that there will be five sets of activities that will create them.

The stages in the life-cycle illustrated in Fig. 1.3 above are described in a little more detail below:

Requirements analysis. This is finding out in detail what the users require of the system that the project is to implement. Some work along these lines will almost certainly have been carried out when the project was evaluated but now the original information obtained needs to be updated and supplemented. Several different approaches to the users' requirements may be explored. For example, a small system that satisfies some, but not all, of the users' needs at a low price may be compared to a system with more functions but at a higher price.

Specification. Detailed documentation of what the proposed system is to do.

Design. A design that meets the specification has to be drawn up. This design activity will be in two stages. One will be the external or user design. This lays down what the system is to look like to the users in terms of menus, screen and report layouts and so on. The next stage produces the physical design, which tackles the way in which the data and software procedures are be structured internally.

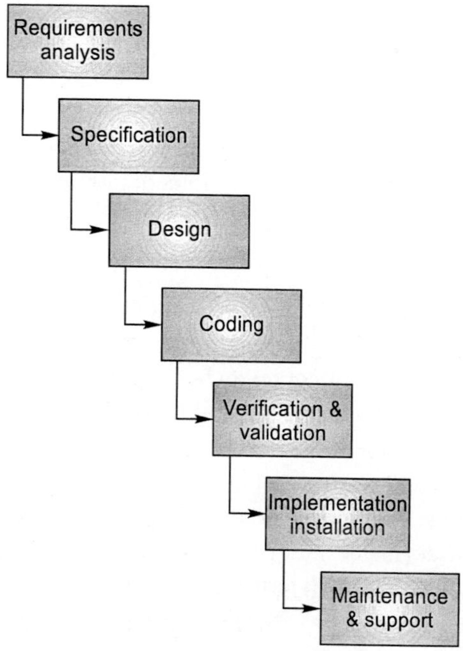

Fig. 1.3 A typical project life-cycle.

Coding. This might refer to writing code in a procedural language such as C or Ada, or might refer to the use of a high level application builder. Even where software is not being built from scratch, some modification to the base application might be required to meet the needs of the new application.

Verification and validation. Whether software is developed specially for the current application or not, careful testing will be needed to check that the proposed system meets its requirements.

Implementation/installation. Some system development practitioners refer to the whole of the project after design as 'implementation' (that is, the implementation of the design) while others insist that the term refers to the installation of the system after the software has been developed. In this case it encompasses such things as setting up data files and system parameters, writing user manuals and training users of the new system.

Maintenance and support. Once the system has been implemented there will be a continuing need for the correction of any errors that may have crept into the system and for extensions and improvements to the system. Maintenance and support activities may be seen as a series of minor software projects. In many environments, most software development is in fact maintenance.

1.8 SOME WAYS OF CATEGORIZING SOFTWARE PROJECTS *(Very Important)*

It is important to distinguish between the main types of software project because what is appropriate in one context might not be so in another. For example, SSADM, the Structured Systems Analysis and Design Method, is suitable for developing information systems but not necessarily other types of system.

Information systems versus embedded systems

A distinction may be made between *information systems* and *embedded systems.* Very crudely, the difference is that in the former case the system interfaces with the organization, whereas in the latter case the system interfaces with a machine. A stock control system would be an information system that controls when the organization reorders stock. An embedded, or process control, system might control the air conditioning equipment in a building. Some systems may have elements of both so that the stock control system might also control an automated warehouse.

Objectives versus products ***(Important)***

Projects may be distinguished by whether their aim is to produce a *product* or to meet certain *objectives.*

A project might be to create a product the details of which have been specified by the client. The client has the responsibility for justifying the product.

On the other hand, the project might be required to meet certain objectives. There might be several ways of achieving these objectives in contrast to the constraints of the product-driven project. One example of this is where a new information system is implemented to improve some service to users inside or outside an organization. The subject of an agreement would be the level of service rather than the characteristics of a particular information system.

Many software projects have two stages. The first stage is an objectives-driven project, which results in a recommended course of action and may even specify a new software application to meet identified requirements. The next stage is a project actually to create the software product.

1.9 THE PROJECT AS A SYSTEM ***(MDU 2010-11, 2012-13) (08 Marks)***

A project is concerned with creating a new system and/or transforming an old one and is itself a system.

Systems, subsystems and environments

A simple definition of the term system is 'a set of interrelated parts'. A system will normally be part of a larger system and will itself comprise *subsystems.*

Outside the system there will be the system's *environment.* This will be made up of things that can affect the system but over which the system has no direct control.

Open versus closed systems

Open systems are those that interact with the environment. Nearly all systems are open. One reason that engineered systems and the projects to construct them often fail is that the technical staff involved do not appreciate the extent to which systems are open and are liable to be affected by outside changes.

Sub-optimization

This is where a subsystem is working at its optimum but is having a detrimental effect on the overall system. An example of this might be where software developers deliver to the users a system that is very efficient in its use of machine resources, but is also very difficult to modify.

Sociotechnical systems

Software projects belong to this category of systems. Any software project requires both technological organization and also the organization of people. Software project managers therefore need to have both technical competence and the ability to interact persuasively with other people.

1.10 PROBLEMS WITH SOFTWARE PROJECTS

One way of deciding what ought to be covered in 'software project management' is to consider what problems need to be addressed.

Traditionally, management has been seen as the preserve of a distinct class within the organization. As technology has made the tasks undertaken by an organization more sophisticated, many management tasks seem to have become dispersed throughout the organization: there are management systems rather than managers. Nevertheless, the successful project will normally have one person who is responsible for its success. Such people are likely to be concerned with the key areas that are most likely to prevent success—they are primarily trouble-shooters and their job is likely to be moulded by the problems that confront the project. A survey of managers published by Thayer, Pyster and Wood identified the following commonly experienced problems:

- poor estimates and plans;
- lack of quality standards and measures;
- lack of guidance about making organizational decisions;
- lack of techniques to make progress visible;
- poor role definition—who does what?
- incorrect success criteria.

The above list looks at the project from the manager's point of view. What about the staff who make up the members of the project team? Below is a list of the problems identified by a number of students on a degree course in Computing and Information Systems who had just completed a year's industrial placement:

- inadequate specification of work;
- management ignorance of IT;
- lack of knowledge of application area;
- lack of standards;
- lack of up-to-date documentation;
- preceding activities not completed on time—including late delivery of equipment;
- lack of communication between users and technicians;
- lack of communication leading to duplication of work;
- lack of commitment—especially when a project is tied to one person who then moves;
- narrow scope of technical expertise;
- changing statutory requirements;
- changing software environment;
- deadline pressure;
- lack of quality control;
- remote management:
- lack of training.

Note how many of the problems identified by the students stemmed from poor communications. Another common problem identified by this and other groups of students is the wide range of IT specialisms–an organization may be made up of lots of individuals or groups who will be expert in one set of software techniques and tools but ignorant of those used by their colleagues. Communication problems are therefore bound to arise.

What about the problems faced by the customers of the products of computet projects? Here are some recent stories in the press:

- the United States Internal Revenue System was to abandon its tax system modernization programme after having spent $4 billion;
- the state of California spent $1 billion on its non-functional welfare database system;
- the £339 million United Kingdom air traffic control system was reported as being two years behind schedule;
- a discount stock brokerage company had 50 people working 14 hours or more a day to correct three months of records clerically—the report commented that the new system had been rushed into operation without adequate testing.
- in the United Kingdom, a Home Office immigration service computerization project was reported as having missed two deadlines and was nine months late;
- the Public Accounts Committee of the House of Commons in the United Kingdom blamed software bugs and management errors for £12 million of project costs in relation to an implementation of a Ministry of Agriculture computer system to administer farm subsidies.

Most of the stories above relate to public sector organizations. This may be misleading—private sector organizations tend to conceal their disasters and in any case many of the public projects above were actually being carried out by private sector contractors. Any lingering faith by users in the innate ability of IT people to plan ahead properly will have been removed by consideration of the 'millennium bug', a purely self-inflicted IT problem. On balance it might be a good idea not to survey users about their problems with IT projects!

1.11 MANAGEMENT CONTROL *(MDU 2008-09, 2011-12) (10 Marks)*

The project control cycle

Management, in general, can be seen as the process of setting objectives for a system and then monitoring the system to see what its true performance is. In Fig. 1.4 the 'real world' is shown as being rather formless. Especially in the case of large undertakings, there will be a lot going on about which management should be aware. As an example, take an IT project that is to replace locally held paper-based records with a centrally-organized database. It might be that staff in a large number of offices that are geographically dispersed need training and then need to use the new IT system to set up the back-log of manual records on the new database It might be that the system cannot be properly operational until the last record has been transferred. It might also be the case that the new system will be successful only if new transactions can be processed within certain time cycles. The managers of the project ought to be asking questions about such things as how effective training has been, how many records have still to be transferred to the new database and transfer rates. This will involve the local managers in *data collection*. Bare details, such as 'location X has processed 2000 documents' will not be very useful to higher management: *data processing* will be needed to transform this *raw data* into useful *information*.

This might be in such forms as 'percentage of records processed', 'average documents processed per day per person' and 'estimated completion date'.

In the example above, the project leader might examine the 'estimated completion date' for completing data transfer for each branch and compare this with the overall target date for completion of this phase of the project. In effect they are comparing actual performance with one aspect of the overall project objectives. They might find that one or two branches are not going to complete the transfer of details in time, and would then need to consider what to do (this is represented in Fig. 1.4 by the box *making decisions/plans*). One possibility would be to move staff temporarily from one branch to another. If this is done, there is always the danger that while the completion date for the one branch is pulled back to before the overall target date, the date for the branch from which staff are being moved is pushed forward beyond that date. The project manager would need to calculate carefully what the impact would be in moving staff from particular branches. This is *modelling* the consequences of a potential solution.

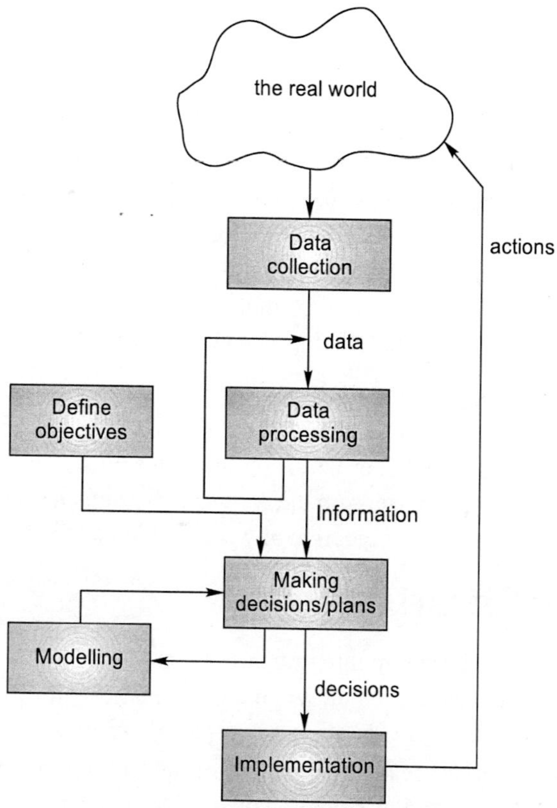

Fig. 1.4 The project control cycle.

Several different proposals could be modelled in this way before one was chosen for *implementation.*

Having implemented the decision, the situation needs to be kept under review by collecting and processing further progress details. For instance, the next time that progress is reported, a branch to which staff have been transferred might still be behind in transferring details. This might be because the reason why the branch has got behind in transferring details is because the manual records are incomplete and another department, for whom the project has a low priority, has to be involved in providing the missing information. In this case, transferring extra staff to do data input will not have accelerated data transfer.

Objectives

To have a successful software project, the manager and the project team members must know what will constitute success. This will make them concentrate on what is essential to project success. Projects objectives should be clearly defined.

There might be more than one set of users of a system and there might be different groups of staff who are involved its development. There is a need for well defined objectives that are accepted by all these people. Where there is more than one user group, then a *project authority* needs to be identified Such a project authority has overall authority over what the project is to achieve.

This authority is often held by a *project steering committee*, which has overall responsibility for setting, monitoring and modifying objectives. The project manager still has responsibility for running the project on a day to day basis, but has to report to the steering committee at regular intervals. Only the steering committee can authorize changes to the project objectives and resources.

Measures of effectiveness

Effective objectives are concrete and well defined. Vague aspirations such as 'to improve customer relations' are unsatisfactory. Objectives should be such that it is obvious to all whether the project has been successful or not. Ideally there should be *measures of effectiveness*, which tell us how successful the project has been. For example, 'to reduce customer complaints by 50%' would be more satisfactory as an objective than 'to improve customer relations.'

Sub-objectives and goals

In order to keep things manageable, objectives might need to be broken down into sub-objectives. Here we say that in order to achieve A we must achieve B, C and D first. These sub-objectives are also known as *goals,* steps on the way to achieving an objective, just as goals scored in a football match are steps towards the objective of winning the match.

1.12 STAKEHOLDERS

(MDU 2008-09) (05 Marks)

Stakeholders are an integral part of a project. They are the end-users or clients, the people from whom requirements will be drawn, the people who will influence the design and ultimately, the people who will reap the benefits of your completed project.

These are people who have a stake or interest in the project. It is important that they be identified as early as possible, because you need to set up adequate communication channels with them right from the start. The project leader also has in be aware that not everybody who is involved with a project has the same motivation and objectives. The end users might, for instance, be concerned about the ease of use of the system while their managers might be interested in the staff savings the new system will allow.

Project stakeholders are individuals and organizations that are actively involved in the project, or whose interests may be affected as a result of project execution or project completion.

Stakeholders might be internal to the project team, external in the project team but in the same organization, or totally external to the organization.

- **Internal to the project team.** This means that they will be under the direct managerial control of the project leader.
- **External to the project team but within the same organization.** For example, the project leader might need the assistance of the information management group in order to add some additional data types to a database or the assistance of the users to carry out systems testing. Here the commitment of the people involved has to be negotiated.
- **External to both the project team and the organization.** External stakeholders might be customers (or users) who will benefit from the system that the project implements or contractors who will carry out work for the project. One feature of the relationship with these people is that it is likely to be based on a legally binding contract.

Within each of the general categories there will be various groups. For example, there will be different types of user with different types of interests.

Different types of stakeholder might have different objectives and one of the jobs of the successful project leader is to recognize these different interests and to be able to reconcile them. It should therefore come as no surprise that the project leader needs to be a good communicator and negotiator. Boehm and Ross proposed a 'Theory W' of software project management where the manager concentrates on creating situations where all parties involved in a project benefit from it and therefore have an interest in its success. (The 'W' stands for Everyone a Winner.)

Important Note : *W^5HH (WHAT, WHY, WHERE, WHEN, WHO, HOW, HOW MUCH) Principle*

W^5HH is an approach suggested by Boehm that addresses all the basic project objectives milestones and schedules, responsibilities, management and technical approaches and required resources. The principles suggested by Boehm's are applicable regardless of the size or complexity of a software project.

The questions noted provide an excellent planning outline for the project managers and the software team.

1.13 REQUIREMENT SPECIFICATION (*MDU 2007-08, 2009-10, 2012-13, 2013-14*) (*10 Marks*)

Very often, especially in the case of product-driven projects, the objectives of the project are carefully defined in terms of functional requirements, quality requirements, and resource requirements.

- **Functional requirements.** These define what the system that will be the end product of the project is to do. Systems analysis and design methods, such as SADT and Information Engineering, are designed primarily to provide functional requirements.
- **Quality requirements.** There will be other attributes of the system to be implemented that do not relate so much to what the system is to do but how it is to do it. These are still things that the user will be able to experience. They include, for example, response time, the ease of using the system and its reliability.
- **Resource requirements.** A record of how much the organization is willing to spend on the system. There will usually be a trade-off between this and the time it takes to implement the system. In general it costs disproportionately more to implement a system by an earlier date than a later one. There might also be a trade-off between the functional and quality requirements and cost. We would all like exceptionally reliable and user-friendly systems that do exactly what we want but we might not be able to afford them.

1.14 INFORMATION AND CONTROL IN ORGANIZATIONS

(*MDU 2009-10, 2012-13*) (*10 Marks*)

Hierarchical information and control systems

With small projects, the project leaders are likely to be working very closely with the other team members and might even be carrying out many non-managerial tasks themselves. Therefore they should have a pretty good idea of what is going on. When projects are larger, many separate teams will be working on different aspects of the project and the overall managers of the project are not going to have day-to-day direct contact with all aspects of the work.

Larger projects are likely to have a hierarchical management structure (Fig. 1.5). Project team members will each have a group leader who allocates them work and to whom they report progress. In turn the group leader, along with several other group leaders, will report to a manager at the next higher level. That manager might have to report to a another manager at a higher level, and so on.

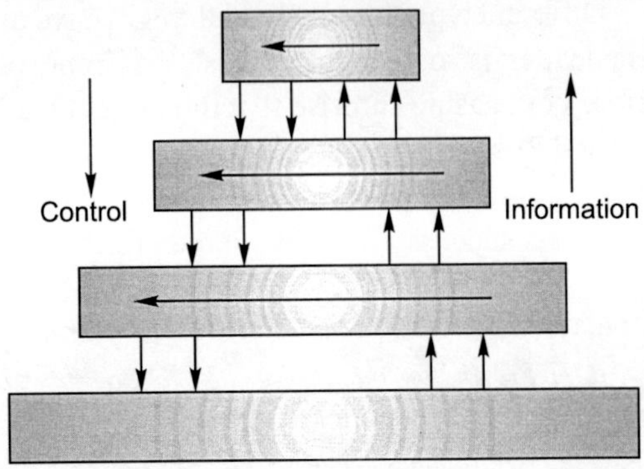

Fig. 1.5 Management information flows up the organizational structure and control flows down.

There might be problems that cannot be resolved at a particular level. For example, additional resources might be needed for some task, or there might be a disagreement with another group. These will be referred to the next higher level of management.

At each higher level more information will be received by fewer people. There is thus a very real danger that managers at the higher levels might be overloaded with too much information. To avoid this, at each level the information will have to be summarized.

As a result of examining the progress information and comparing it against what was planned, some remedial action might need to be taken. Instructions may be formulated and passed down to a lower level of management. The lower level managers will have to interpret what needs to be done and formulate more detailed plans to fulfil the directive. As the directives filter down the hierarchy, they will be expanded into more detail at each level.

Not all information flows concerning a project will be going up and down the hierarchy. There will also be lateral flows between groups and individuals on the same level.

For example, a programmer will receive a specification from an analyst and might then seek clarification.

Levels of decision making and information

Each decision made in a project environment should be based on adequate information of the correct sort. The type of information needed depends on the level of decision making. Decisions can be grouped at three levels: *Strategic, tactial* and *operational.*

Strategic decision making is essentially about deciding objectives. In the case of the Brightmouth College payroll, the decision to become administratively independent could be regarded as a strategic decision. In our example we were interested only in the payroll, but this might have been part of a wider programme which may have affected many other administrative functions.

Tactical decision making is needed to ensure that the objectives will be fulfilled. The project leader who has the responsibility for achieving objectives will have to formulate a plan of action to meet those objectives. The project leader will need to monitor progress to see whether these objectives are likely to be met and to take action where needed to ensure that the things remain on course.

Operational decisions relate to the day-to-day work of implementing the project. Deciding the content of the acceptance tests might come under this heading.

Differences in types of information

Table 1.1 gives some idea of the differences in the kind of information needed. There is a kind of continuum for most of the qualities suggested and what is needed for tactical decision making comes somewhere in the middle.

Effectiveness is concerned with doing the right thing. *Efficiency* is carrying out a task making the best possible use of the resources.

Measurement

The leader of a small project will have direct contact with many aspects of the project. With larger projects, project leaders would have to depend on information being supplied to them. This information should not be vague and ideally should be quantitative. This lies in with our need for unambiguous measures of effectiveness.

Software development deals largely with intangibles and does not easily lend itself to quantitative measures, but attempts are increasingly being made to introduce measurement into the software process.

Table 1.1: The types of information required for decision making

Characteristic	*Operational*	*Strategic*
motivation	efficiency	effectiveness
orientation	internal	internal and external
focus	specific to a function	specific to organization
detail	detailed	summarized
response	fast	not so fast
access paths	standard	flexible
up-to-dateness	essential	desirable
accuracy	essential	approximate
certainty	essential	often predictive
objectivity	high	more subjective
information type	mainly quantitative	often qualitative

Software measurements can be divided into *performance measures* and *predictive measures.*

- **Performance measures.** These measure the characteristics of a system that has been delivered. They are important when we are trying to specify unambiguously the quality requirements of a proposed system.
- **Predictive measures.** The trouble with performance measures is that you need to have a system actually up and running before you can take measurements. As a project leader, what you want to be able to do is to get some idea of the likely characteristics of the final system during its development. *Predictive measures* are taken during development and indicate what the performance of the final system is likely to be.

For example, the errors found per KLOC (that is, thousand lines of code) at different stages of the project might help to predict the correctness and reliability of the final system. Keystrokes required to carry out a particular transaction might help predict what the operator time to carry out the transaction is likely to be. Modularity, the degree to which the software is composed of self-contained manageable components, helps predict how easy changes to the final system will be.

CONCEPTUAL SHORT QUESTIONS WITH ANSWERS

1. List down the steps that will be followed for network-based project management methodology.

Ans. The main steps of network based project management methodology are as follows:

Step 1: Project Planning

Step 2: Time and Resource Estimation

Step 3: Basic Scheduling Calculations

Step 4: Time-Cost Trade-offs

Step 5: Resources Allocation

Step 6: Project Control.

2. PFD stands for?

Ans. PFD stands for Product Flow Diagrams.

3. CASE stands for?

Ans. CASE stands for Computer Aided Software Engineering.

4. List at least seven deliverables in a software project.

Ans. SRS document, SDD, Source-Code Object-Code, Test Plan, Quality Plan and Project Plan.

5. Project Management involves four major activities. What are they?

Ans. (*a*) Feasibility Study (*b*) Project Planning

(*c*) Project Execution (*d*) Project Termination.

REVIEW QUESTIONS

1. What do you understand by the term 'Software Project Management'? What are the control problems in producing the requirements specifications for a system?

2. Define the term Software Project. How is it different from other project? Describe the various activities involved in software project design and implementation.

3. You are given a project to develop a website for the college. Determine the scope, design skill for development.

4. Comment on the statement:

(*a*) "Unambiguous requirement specification is an essential requirement for success of the project."

(*b*) "Software projects may seem similar to other engineering projects but have some attributes which make them different".

5. Project management involves four major activities. What are they?

6. "Project management is a conversion process." Comment.

7. Explain the following qualities of the requirement specification:

(*a*) Correct and ambiguous (*b*) Verifiable and traceable

(*c*) Compute and consistent

8. Would an operating system on a computer be an information system or an embedded system?

9. Define the specific uses of Project Management.

10. Define the SPM framework with example.

11. Define the various tips for avoiding project failure.

12. Define the term Management Control with suitable example.

13. What are the characteristics of a Project Manager?

14. Explain the role and responsibilities of a Project Manager?

CHAPTER 2

STEP WISE PROJECT PLANNING

INSIDE THIS CHAPTER

2.1 INTRODUCTION TO STEP WISE PROJECT PLANNING

(MDU 2008-09, 2010-11, 2012-13) *(20 Marks)*

Software project management begins with a set of activities that are collectively called project planning. The purpose of Software Project Planning is to establish reasonable plans for performing all activities that are the part of Software Engineering. Project Planning is a discipline for stating how to complete a project within a certain timeframe, usually with defined stages, and with designated resources. This chapter describes a framework of basic steps in project planning and control upon which the following chapters build. There are many different techniques that can be used in project planning and this chapter gives an overview of the points at which these techniques can be used during project planning.

Project planning is a part of project management, which relates to the use of schedules such as Gantt Charts to plan and subsequently report progress within the project enviornment.

The Software Project Management Plan (SPMP) is the controlling document for a software project. It specifies the technical and managerial approaches to be used to develop software components. For example, SPMP may include configuration management, quality assurance, verification and validation.

An SPMP is the project manager's statement of understanding with the development organisations and the project team members.

The framework described is called the Step Wise method to help to distinguish it from other methods such as PRINCE 2. PRINCE 2 is the set of project management standards that have been published by the Central Computing and Telecommunications Agency (CCTA) for use on British government IT projects. The standards are also widely used on non-government projects in the United Kingdom. Step Wise should be compatible with PRINCE 2. It should be noted, however, that Step Wise covers only the planning stages of a project and not monitoring and control.

In order to illustrate the Step Wise approach and to show how it might have to be adapted to deal with different circumstances, two parallel examples are used.

> **Important Note:** *There is a general rule, "There can be no project deliverables that do not have activities that create them". The reverse is also true. It states that there are no activities that do not produce project deliverables.*

In Table 2.1 we outline the general approach that might be taken to planning these projects. Fig. 2.1 provides an outline of the main planning activities. Steps 1 and 2, 'Identify project scope and objectives' and 'Identify project infrastructure', may be tackled in parallel in some cases. Steps 5 and 6 will have to be repeated for each activity needed to complete the project.

A major principle of project planning is to plan in outline first and then in more detail as the time to carry out an activity approaches. Hence the lists of products and activities that are the result of Step 4 will the reviewed when the tasks connected with a particular phase of a project are considered in more detail. This will be followed by a more detailed iteration of Steps 5 to 8 for the phase under consideration.

2.2 STEP 0: SELECT PROJECT

This is called Step 0 because in a way it is outside the main project planning process. Projects are not initiated out of thin air—some activity has to take place before deciding that this project rather than another is worth undertaking. This project evaluation may be done on an individual basis or as part of strategic planning.

Table 2.1: An outline of Step Wise planning activities

Step	*Activities within step*
0	Select project
1	Identify project scope and objectives
	1.1 Identify objectives and measures of effectiveness in meeting them
	1.2 Establish a project authority
	1.3 Identify all stakeholders in the project and their interests
	1.4 Modify objectives in the light of stakeholder analysis
	1.5 Establish methods of communications with all parties
2	Identify project infrastructure
	2.1 Establish relationship between project and strategic planning
	2.2 Identity installation standards and procedures
	2.3 Identity project team organization
3	Analyse project characteristics
	3.1 Distinguish the project as either objective-or product-driven
	3.2 Analyse other project characteristics
	3.3 Identify high level project risks
	3.4 Take into account user requirements concerning implementation
	3.5 Select general lifecycle approach
	3.6 Review overall resource estimates

Step	*Activities within step*
4	Identify project products and activities 4.1 Identify and describe project products (or deliverables) 4.2 Document generic product flows 4.3 Recognize product instances 4.4 Produce ideal activity network 4.5 Modify ideal to take into account need for stages and checkpoints
5	Estimate effort for each activity 5.1 Carry out bottom-up estimates 5.2 Revise plan to create controllable activities
6	Identify activity risks 6.1 Identify and quantify activity-based risks 6.2 Plan risk reduction and contingency measures where appropriate 6.3 Adjust plans and estimates to take account of risks
7	Allocate resources 7.1 Identify and allocate resources 7.2 Revise plans and estimates to account for resource constraints
8	Review/publicize plan 8.1 Review quality aspects of project plan 8.2 Document plans and obtain agreement
9.	Execute plan
10.	Lower levels of planning

2.3 STEP 1: IDENTIFY PROJECT SCOPE AND OBJECTIVES

The activities in this step ensure that all the parties to the project agree on the objectives and are committed to the success of the project. A danger to be avoided is overlooking people who are affected by the project.

Step 1.1: Identify objectives and practical measures of the effectiveness in meeting those objectives

We discussed earlier the need for agreed objectives for a project and ways of measuring the success in achieving those objectives.

Step 1.2: Establish a project authority

A single overall project authority needs to be established so that there is unity of purpose among all those concerned.

Step 1.3: Identify all stakeholders in the project and their interests

Recall that this was the basis of a discussion in Chapter 1. Essentially all the parties who have an interest in the project need to be identified.

Step 1.4: Modify objectives in the light of stakeholder analysis

In order to gain the full cooperation of all concerned, it might be necessary to modify the project objectives. This can mean adding new features to the system giving a benefit to some stakeholder group as a means of assuring their commitment to the project. This is potentially dangerous, since the system size might be increased and the original objectives obscured. Because of these dangers, this process must be done consciously and in a controlled manner.

Step 1.5: Establish methods of communication with all parties

For internal staff, this should be fairly straightforward, but a project leader implementing a payroll system would need to find a contact point with BACS (Bankers Automated Clearing Scheme) for instance.

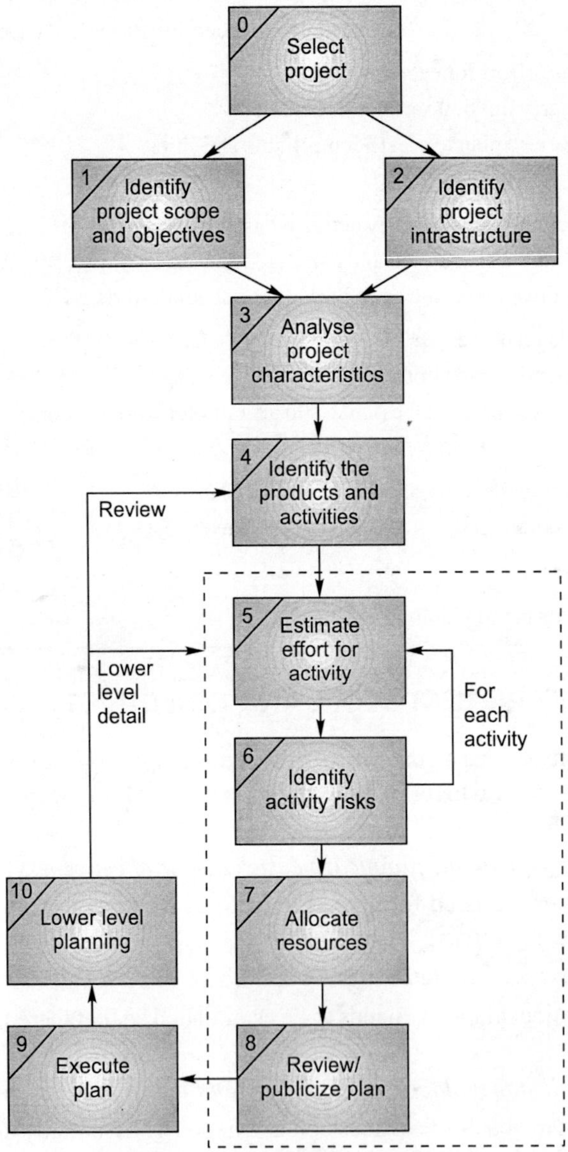

Fig. 2.1 An overview of step wise.

2.4 STEP 2: IDENTIFY PROJECT INFRASTRUCTURE

Projects are rarely initiated in a vacuum. There is usually some kind of existing infrastructure into which the project can fit. The project leader who does not already know about this structure needs to find out its precise nature.

Step 2.1: Identify relationship between the project and strategic planning

As well as identifying projects to be carried out, an organization needs to decide the order in which these projects are to be carried out. It also needs to establish the framework within which the proposed new systems are to fit. Hardware and software standards, for example, are needed so that various systems can communicate with each other. These strategic decisions must be documented in a strategic business plan or in an information technology plan that is developed from the business plan.

Step 2.2: Identify installation standards and procedures

Any organization that develops software should define its development procedures. As a minimum, the normal stages in the software life cycle to be carried out should be documented along with the products created at each stage.

Change control and *configuration management* standards should be in place to ensure that changes to requirements are implemented in a safe and orderly way.

The procedural standards may lay down the quality checks that need to be done at each point of the project life cycle or these may be documented in a separate *quality standards and procedures* manual.

The organization, as part of its monitoring and control policy must have in place a *measurement programme* that dictates that certain statistics have to be collected at various stages of a project.

Finally the project manager should be aware of any *project planning and control standards.* These will relate to the way that the project is controlled: for example, the way that the hours spent by team members on individual tasks are recorded on time-sheets.

Step 2.3: Identify project team organization

Project leaders, especially in the case of large projects, will often have some control over the organizational structure of the project team. More often, though, the organizational structure will be dictated to them. For example, there might have been a high level managerial decision that code developers and systems analysts will be in different groups, or that the development of PC applications will not be done within the same group as that responsible for 'legacy' mainframe applications.

If the project leader does have some control over the project team organization then this would best be considered at a later stage (see Step 7: Allocate resources).

2.5 STEP 3: ANALYSE PROJECT CHARACTERISTICS

The general purpose of this part of the planning operation is to ensure that the appropriate methods are used for the project.

Step 3.1: Distinguish the project as either objective- or product-driven

This has already been discussed in the first chapter. A general point to note is that as system development advances, it tends to become more product-driven, although the underlying objectives always remain and must be respected.

Step 3.2: Analyse other characteristics (including quality-based ones)

For example, is this an information system that is being developed or a process control system, or does it have elements of both? Is it a safety-critical system, that is, where human life could be threatened by a malfunction?

Step 3.3: Identify high level project risks

Consideration must be given to the risks that threaten the successful outcome of the project. Generally speaking most risks can be attributed to the operational or development environment, the technical nature of the project or the type of product being created.

Step 3.4: Take into account user requirements concerning implementation

The clients will usually have their own procedural requirements. For example, work for government departments usually requires the use of SSADM.

Step 3.5: Select general lifecycle approach in the light of the above

The project life cycle to be used for the project will be influenced by the issues raised above. For example, a prototyping approach might be used where the user requirements are not clear.

Step 3.6: Review overall resource estimates

Once the major risks have been identified and the broad project approach has been decided upon, this would be a good point at which to reestimate the effort and other resources required to implement the project. Where enough information is available, an estimate based on function points might be appropriate.

2.6 STEP 4: IDENTIFY PROJECT PRODUCTS AND ACTIVITIES

The more detailed planning of the individual activities that will be needed now takes place. The longer term planning is broad and in outline, while the more immediate tasks are planned in some detail.

Step 4.1: Identify and describe project products (or deliverables)

In general there can be no project products that do not have activities that create them. Wherever possible, we ought also to ensure the reverse: that there are no activities that do not produce a tangible product. Making sure we have identified all the things the project is to create helps us to ensure that all the activities we need to carry out are accounted for.

These products will include a large number of *technical* products including training material and operating instructions, but also products to do with the *management* and the *quality* of the project. Planning documents would, for example, be management products.

The products will form a hierarchy. The main products will have sets of component products, which in turn might have sub-component products and so on. These relationships can be documented in a Product Breakdown Structure (PBS).

This part of the planning process draws heavily on the standards laid down in Prince 2. These specify that products at the bottom of the PBS should be documented by *Product Descriptions*, which contain:

- the name/identity of the product;
- the purpose of the product;
- the derivation of the product (that is, the other products from which it is derived);
- the composition of the product;
- the form of the product;

- the relevant standards;
- the quality criteria that should apply to it.

Step 4.2: Document generic product flows

Some of the products will need some other product to exist first before they can be created. For example, a program design must be created before the program can be written and the program specification must exist before the design can be commenced. These relationships can be portrayed in a Product Flow Diagram (PFD). Fig. 2.2 gives an example.

> **Important Note:** *From your Product Flow Diagram (PFD) also, you can draw an activity diagram.*

Step 4.3: Recognize product instances

Where the same generic PFD fragment relates to more than one instance of a particular type of product, an attempt should be made to identify each of those instances.

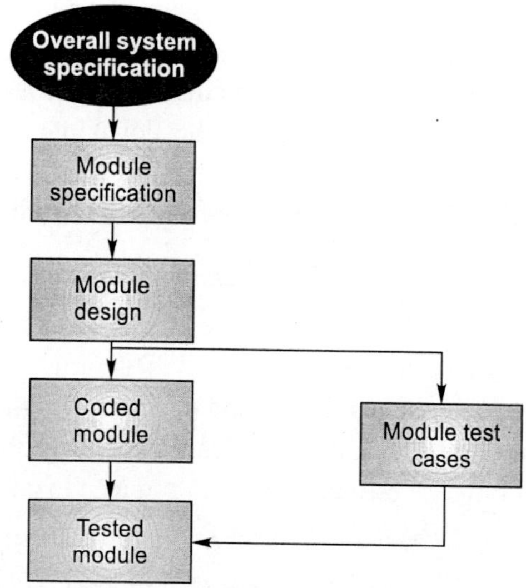

Fig. 2.2 A fragment of a PFD.

Step 4.4: Produce ideal activity network

In order to generate one product from another there must be one or more activities that carry out the transformation. By identifying these activities we can create an activity network, which shows the tasks that have to be carried out and the order in which they have to be executed.

The activity networks are 'ideal' in the sense that no account has been taken of resource constraints.

Step 4.5: Modify the ideal to take into account need for stages and checkpoints

The approach to sequencing activities described above encourages the formulation of a plan that will minimize the overall duration, of 'elapsed time', for the project. It assumes that an activity will start as soon as the preceding ones upon which it depends have been completed.

There might, however, be a need to modify this by dividing the project into stages and introducing checkpoint activities. These are activities that draw together the products of preceding activities to check that they are compatible. These checkpoints are sometimes referred to as milestone events. A checkpoint could potentially delay work on some elements of the project – there has to be a trade-off between efficiency and quality.

2.7 STEP 5: ESTIMATE EFFORT FOR EACH ACTIVITY

Step 5.1: Carry out bottom-up estimates

Some top-down estimates of effort, cost and duration will already have been done (see Step 3.6).

At this point, estimates of the staff effort and other resources required, and the probable elapsed time needed for each activity will need to be produced. The method of arriving at each of these estimates will vary depending on the type of activity.

The individual activity estimates of effort should be summed to get an overall bottom-up estimate, which can be reconciled with the previous top-down estimate.

The activities on the activity network can be annotated with their elapsed times so that the overall duration of the project can be calculated.

Step 5.2: Revise plan to create controllable activities

The estimates for individual activities might reveal that some are going to take quite a long time. Long activities often make a project difficult to control. If an activity involving system testing is to take 12 weeks, it might be difficult after six weeks to judge accurately whether 50% of the work is completed. It would be better to break this down into a series of smaller sub-tasks.

2.8 STEP 6: IDENTIFY ACTIVITY RISKS

Step 6.1: Identify and quantify activity-based risks

Risks inherent in the overall nature of the project have already been considered in Step 3. We now want to look at each activity in turn and assess the risks to its successful outcome. The seriousness of each risk and likelihood of it occurring have to be gauged. At individual task level some risks are unavoidable, and the general effect if a problem materializes is to make the task longer or more costly. A range of estimates can be produced to take into account the possible occurrence of the risks.

Step 6.2: Plan risk reduction and contingency measures where appropriate

It is possible to avoid or at least reduce some of the identified risks. Contingency plans specify action that is to be taken if a risk materializes. For example, a contingency plan could be to use contract staff if a member of the project team is unavailable at a key time because of illness.

Step 6.3: Adjust overall plans and estimates to take account of risks

We can change our plans, perhaps by adding new activities which reduce risks. For example, a new programming language could mean that we schedule training courses and time for the programmers to practise their new programming skills on some non-essential work.

2.9 STEP 7: ALLOCATE RESOURCES ***(Important)***

Step 7.1: Identify and allocate resources

The type of staff needed for each activity is recorded. The staff available for the project are identified and are provisionally allocated to tasks.

Step 7.2: Revise plans and estimates to take into account resource constraints

Some staff might be needed for more than one task at the same time and, in this case, an order of priority is established. The decisions made here can have an effect on the overall duration of the project when some tasks are delayed while waiting for staff to become free.

Ensuring someone is available to start work on an activity as soon as the proceeding activities have been completed might mean that they are idle while waiting for the job to start and are therefore used inefficiently.

2.10 STEP 8: REVIEW/PUBLICIZE PLAN

Step 8.1: Review quality aspects of the project plan

A danger when controlling any project is that an activity can reveal that an earlier activity was not properly completed and needs to be reworked. This, at a stroke, can transform a project that appears to be progressing satisfactorily into one that is badly out of control. It is important to know that when a task is reported as completed, it really is, hence the importance of quality reviews. Each task should have "exit requirements'. These are quality cheeks that have to be passed before the activity can be 'signed off' as completed.

Step 8.2: Document plans and obtain agreement

It is important that the plans be carefully documented and that all the parties to the project understand and agree to the commitments required of them in the plan.

2.11 STEPS 9 AND 10: EXECUTE PLAN AND LOWER LEVELS OF PLANNING

Once the project is under way, plans will need to be drawn up in greater detail for each activity as it becomes due. Detailed planning of the later stages will have to be delayed because more information will be available nearer the start of the stage. Of course, it is necessary to make provisional plans for the more distant tasks, because thinking about what has to be done can help unearth potential problems, but sight should not be lost of the fact that these plans are provisional.

CONCEPTUAL SHORT QUESTIONS WITH ANSWERS

1. What is Parkinson's Law?

Ans. According to Parkinson's Law, work expands to fill the available time and budget.

2. SPMP stands for?

Ans. SPMP stands for Software Project Management Plan.

3. PDF stands for?

Ans. PDF stands for Product Flow Diagrams.

4. What is Slip Chart?

Ans. It show those activities that are not progressing to schedule. The more the slip line bends, the greater is the variation from the plan. Additional slip lines are added at intervals. A jagged slip line indicates a need for rescheduling.

REVIEW QUESTIONS

1. What are the activities involved in the planning of a software project? Discuss them.
2. Write a short note on software project planning discussing its all aspects.
3. Describe any software project management software which you have studied or worked on.
4. Describe the contents list of project plan.
5. Suggest some approaches or methods to shorter the project duration time.
6. List the products created by step wise planning process.
7. What products must exist before the activity test 'program' can take place? What products does this activity create?
8. Define stepwise project planning with suitable example.

CHAPTER 3

PROJECT EVALUATION AND ESTIMATION

INSIDE THIS CHAPTER

3.1 INTRODUCTION

Deciding whether or not to go ahead with a project is really a case of comparing a proposed project with the alternatives and deciding whether to proceed with it. That evaluation will be based on strategic, technical and economic criteria and will normally be undertaken as part of strategic planning or a feasibility study for any information system development. The risks involved also need to be evaluated.

In this chapter we shall be using the term *project* in a broader sense than elsewhere in the book. Our decision as to whether or not to proceed with a project needs to be based upon whether or not it is desirable to carry out the *development and operation* of a software system. The term project may therefore be used, in this context, to describe the whole life cycle of a system from conception through to final decommissioning.

Project evaluation is normally carried out in Step 0 of Step Wise (Fig. 3.1). The subsequent steps of Step Wise are then concerned with managing the development project that stems from this project selection.

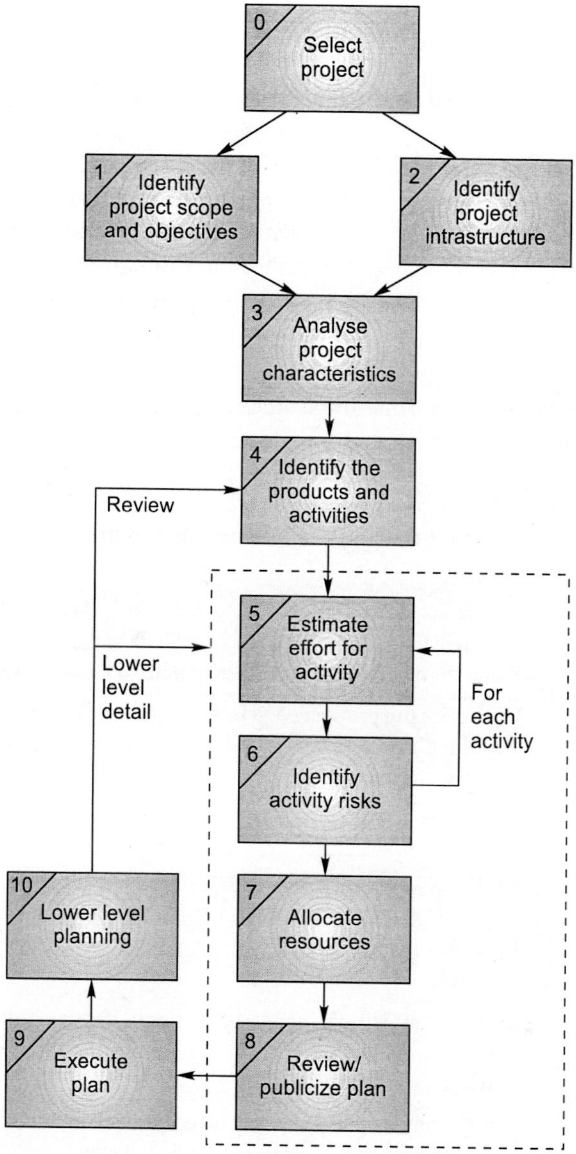

Fig. 3.1 Project evaluation is carried out in step 0.

3.2 STRATEGIC ASSESSMENT

(MDU 2011-12) (10 Marks)

Programme management

It is being increasingly recognized that individual projects need to be seen as components of a *programme* and should be evaluated and managed as such. A programme, in this context, is a collection of projects that all contribute to the same overall organizational goals. Effective programme management requires that there is a well defined *programme goal* and that all the organization's projects are selected and tuned to contribute to this goal. A project must be evaluated according to how it contributes to this programme goal and its viability, timing, resourcing and final worth can be affected by the programme as a whole. It is to be expected that the value of any project is increased by the fact that it is part of a programme, the whole, as they say, being greater than the sum of the parts.

In order to carry out a successful strategic assessment of a potential project there should therefore be a strategic plan clearly defining the organization's objectives. This provides the context for defining the programme and programme goals and, hence, the context for assessing the individual project. It is likely, particularly in a large organization, that there will be an organizational structure for programme management and it will be, for example, the *programme director* and *programme executive,* rather than, say, a project manager, who will be responsible for the strategic assessment of a proposed project.

Even where there is no explicitly defined programme, any proposed project must be evaluated within the context of the organization's overall business objectives. Moreover, any potential software system will form part of the user organization's overall information system and must be evaluated within the context of the existing information system and the organization's information strategy. Table 3.1 illustrates typical issues that must be addressed as part of the strategic assessment of a project.

Table 3.1: Typical issues and questions to be considered during strategic assessment

Issue	*Typical questions*
Objectives	How will the proposed system contribute to the organization's stated objectives? How, for example, might it contribute to an increase in market share?
IS plan	How does the proposed system fit into the IS plan? Which existing system(s) will it replace/interface with? How will it interact with systems proposed for later development?
Organization structure	What effect will the new system have on the existing departmental and organization structure? Will, for example, a new sales order processing system overlap existing sales and stock control functions?
MIS	What information will the system provide and at what levels in the organization? In what ways will it complement or enhance existing management information systems?
Personnel	In what way will the proposed system affect manning levels and the existing employee skill base? What are the implications for the organization's overall policy on staff development?
Image	What, if any, will be the effect on customers' attitudes towards the organization? Will the adoption of, say, automated systems conflict with the objectives of providing a friendly service?

Where a well-defined information systems strategy does not exist, system development and the assessment of project proposals will be based on a more piecemeal approach—each project being individually assessed early in its life cycle. In such cases it is likely that cost-benefit analysis will have more importance and some of the questions of Table 3.1 will be more difficult to answer.

Portfolio management

Where an organization such as a software house is developing a software system they could be asked to carry out a strategic and operational assessment on behalf of the customer. Whether or not this should be the case, they will require an assessment of any proposed project themselves. They will need to ensure that carrying out the development of a system is consistent with their own strategic plan—it is unlikely, for example, that a software house specializing in financial and accounting systems would wish to undertake development of a factory control system unless their strategic plan placed an emphasis on diversification.

The proposed project will form part of a *portfolio* of ongoing and planned projects and the selection of projects must take account of the possible effects on other projects in the portfolio (competition for resources, for example) and the overall portfolio profile (for example, specialization versus diversification).

3.3 TECHNICAL ASSESSMENT

Technical assessment of a proposed system consists of evaluating the required functionality against the hardware and software available. Where an organization has a strategic information systems plan, this is likely to place limitations on the nature of solutions that might be considered. The constraints will, of course, influence the cost of the solution and this must be taken into account in the cost-benefit analysis.

3.4 COST-BENEFIT ANALYSIS

(MDU 2011-12, 2012-13, 2013-14) (10 Marks)

The most common way of carrying out an economic assessment of a proposed information system, or other development, is by comparing the expected costs of development and operation of the system with the benefits of having it in place.

Assessment is based upon the question of whether the estimated costs are exceeded by the estimated income and other benefits. Additionally, it is usually necessary to ask whether or not the project under consideration is the best of a number of options. There might be more candidate projects than can be undertaken at any one time and, in any case, projects will need to be prioritized so that any scarce resources may be allocated effectively.

The standard way of evaluating the economic benefits of any project is to carry out a cost-benefit analysis, which consists of two steps:

- **Identifying and estimating all of the costs and benefits of carrying out the project.** This includes development costs of the system, the operating costs and the benefits that are expected to accrue from the operation of the system. Where the proposed system is replacing an existing one, these estimates should reflect the costs and benefits due to the new system. A sales order processing system, for example, could not claim to benefit an organization by the total value of sales—only by the increase due to the use of the new system.
- **Expressing these costs and benefits in common units.** We must evaluate the net benefit, which is the difference between the total benefit and the total cost. To do this, we must express each cost and each benefit in monetary terms.

Most costs are relatively easy to identify and quantify in approximate monetary terms. It is helpful to categorize costs according to where they originate in the life of the project.

- **Development costs**—include the salaries and other employment costs of the staff involved in the development project and all associated costs.

- **Setup costs**—include the costs of putting the system into place. These consist mainly of the costs of any new hardware and ancillary equipment but will also include costs of file conversion, recruitment and staff training
- **Operational costs** consist of the costs of operating the system once it has been installed.

Benefits, on the other hand, are often quite difficult to quantify in monetary terms even once they have been identified. Benefits may be categorized as follows:

- **Direct benefits**—these accrue directly from the operation of the proposed system. These could, for example, include the reduction in salary bills through the introduction of a new, computerized system.
- **Assessable indirect benefits**—these are generally secondary benefits, such as increased accuracy through the introduction of a more user-friendly screen design where we might be able to estimate the reduction in errors, and hence costs, of the proposed system.
- **Intangible benefits**—these are generally longer term or benefits that are considered very difficult to quantify. Enhanced job interest can lead to reduced staff turnover and, hence, lower recruitment costs.

Any project that shows an excess of benefits over costs is clearly worth considering for implementation. However, as we shall see later, it is not a sufficient justification for going ahead; we might not be able to afford the costs; there might be even better projects we could allocate our resources to instead; the project might be too risky.

3.5 CASH FLOW FORECASTING

As important as estimating the overall costs and benefits of a project is the forecasting of the cash flows that will take place and their timing. A cash flow forecast will indicate when expenditure and income will take place (Fig. 3.2).

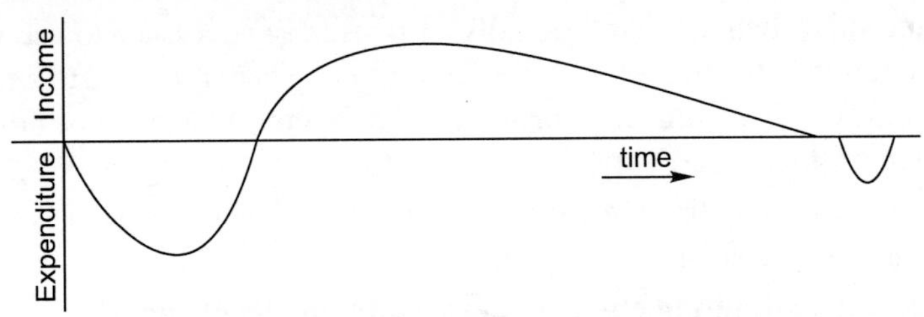

Fig. 3.2 Typical product life cycle cash flow.

We need to spend money, such as staff wages, during the development stages of a project. Such expenditure cannot be deferred until income is received (either from using the software if it is being developed for in-house use or from selling it). It is important that we know that we can fund the development expenditure either from the company's own resources or by borrowing from the bank. In any event, it is vital to have some forecast of when expenditure such as the payment of salaries and bank interest will take place and when any income is to be expected, such as payment on completion or, possibly, stage payments.

Accurate cash flow forecasting is not easy, as it generally needs to be done early in the project's life cycle (at least before any significant expenditure is committed) and many items to be estimated (particularly the benefits of using software or decommissioning costs) might be some years in the future.

When estimating future cash flows, it is usual to ignore the effects of inflation. Trying to forecast the effects of inflation increases the uncertainty of the forecasts. Moreover, if expenditure is increased due to inflation it is likely that income will increase proportionately. However, measures to deal with increases in costs where work is being done for an external customer must be in place, for example, index linked prices where work involves use of raw materials.

Tablc 3.2 illustrates cash flow forecasts for four projects. In each case it is assumed that the cash flows take place at the end of each year. For short-term projects or where candidate projects demonstrate significant seasonal cash flow patterns it can be advisable to produce quarterly, or even monthly, cash flow forecasts.

3.6 COST-BENEFIT EVALUATION TECHNIQUES *(MDU 2012-13, 2013-14) (10 Marks)*

We would consider proceeding with a project only where the benefits outweigh the costs. However, in order to choose among projects, we need to take into account the timing of the costs and benefits as well as the benefits relative to the size of the investment.

In the following sections we will take a brief look at some common methods for comparing projects on the basis of their cash flow forecasts.

Net profit

The net profit of a project is the difference between the total costs and the total income over the life of the project. Project 2 in Table 3.2 shows the greatest net profit but this is at the expense of a large investment. Indeed, if we had £ 1 m to invest, we might undertake all of the other three projects and obtain an even greater net profit. Note also, that all projects contain an element of risk and we might not be prepared to risk £1 m.

Table 3.2: Four project cash flow projections–figures are end of year totals (£)

Year	*Project 1*	*Project 2*	*Project 3*	*Project 4*
0	– 100,000	– 1,000,000	– 100,000	– 120,000
1	10,000	200,000	30,000	30,000
2	10,000	200,000	30,000	30,000
3	10,000	200,000	30,000	30,000
4	20,000	200,000	30,000	30,000
5	100,000	300,000	30,000	75,000
Net profit	50,000	100,000	50,000	75,000

Moreover, the simple net profit takes no account of the timing of the cash flows. Projects 1 and 3 each have a net profit of £50,000 and therefore, according to this selection criterion, would be equally preferable. The bulk of the income occurs late in the life of project 1, whereas project 3 returns a steady income throughout its life. Having to wait for a return has the disadvantage that the investment must be funded for longer. Add to that the fact that, other things being equal, estimates in the more distant future are less reliable that short-term estimates and we can see that the two projects are not equally preferable.

Payback period

The *payback period* is the time taken to break even or pay back the initial investment. Normally, the project with the shortest payback period will be chosen on the basis that an organization will wish to minimize the time that a project is 'in debt'.

The advantage of the payback period is that it is simple to calculate and is not particularly sensitive to small forecasting errors. Its disadvantage as a selection technique is that it ignores the overall profitability of the project–in fact, it totally ignores any income (or expenditure) once the project has broken even. Thus the fact that projects 2 and 4 are, overall, more profitable than project 3 is ignored.

Return on investment *(Important)*

The *return on investment* (ROI), also known as the *accounting rate of return* (ARR), provides a way of comparing the net profitability to the investment required. There are some variations on the formula used to calculate the return on investment but a straightforward common version is

$$\text{ROI} = \frac{\text{average annual profit}}{\text{total investment}} \times 100$$

The return on investment provides a simple, easy to calculate measure of return on capital and is therefore quite popular. Unfortunately it suffers from two severe disadvantages. Like the net profitability, it takes no account of the timing of the cash flows. More importantly, it is tempting to compare the rate of return with current interest rates. However, this rate of return bears no relationship to the interest rates offered or charged by banks (or any other normal interest rate) since it takes no account of the timing of the cash flows or of the compounding of interest. It is therefore, potentially, very misleading.

Net present value

The calculation of *net present value* is a project evaluation technique that lakes into account the profitability of a project and the timing of the cash flows that are produced. It does so by discounting future cash flows by a percentage known as the discount rate. This is based on the view that receiving £100 today is better than having to wait until next year to receive it, because the £ 100 next year is worth less than £100 now. We could, for example, invest the £100 in a bank today and have £100 plus the interest in a year's time. If we say that the present value of £100 in a year's time is £91, we mean that £100 in a year's time is the equivalent of £91 now.

The equivalence of £91 now and £100 in a year's time means we are discounting the future income by approximately 10%, that is, we would need an extra 10% to make it worthwhile waiting for a year. An alternative way of considering the equivalence of the two is to consider that, if we received £91 now and invested for a year at an annual interest rate of 10%, it would be worth £100 in a year's time. The annual rate by which we discount future earnings is known as the *discount rate* 10% in the above example.

Similarly, £100 received in 2 years' time would have a present value of approximately £83, in other words, £83 invested at an interest rate of 10% would yield approximately £100 in 2 years' time.

The present value of any future cash flow may be obtained by applying the following formula

$$\text{Present value} = \frac{\text{value in year } t}{(1+t)^t}$$

where r is the discount rate, expressed as a decimal value and t is the number of years into the future that the cash flow occurs.

Alternatively, and rather more easily, the present value of a cash flow may be calculated by multiplying the cash flow by the appropriate discount factor. A small table of discount factors is given in Table 3.3.

The NPV for a project is obtained by discounting each cash flow (both negative and positive) and summing the discounted values. It is normally assumed that any initial investment takes place immediately (indicated as year 0) and is not discounted. Later cash flows are normally assumed to take place at the end of each year and are discounted by the appropriate amount.

Table 3.3: Table of NPV discount factors

Year	Discount rate (%)					
	5	*6*	*8*	*10*	*12*	*15*
1	0.9524	0.9434	0.9259	0.9091	0.8929	0.8696
2	0.9070	0.8900	0.8573	0.8264	0.7972	0.7561
3	0.8638	0.8396	0.7938	0.7513	0.7118	0.6575
4	0.8227	0.7921	0.7350	0.6830	0.6355	0.5718
5	0.7835	0.7473	0.6806	0.6209	0.5674	0.4972
6	0.7462	0.7050	0.6302	0.5645	0.5066	0.4323
7	0.7107	0.6651	0.5835	0.5132	0.4523	0.3759
8	0.6768	0.6274	0.5403	0.4665	0.4039	0.3269
9	0.6446	0.5919	0.5002	0.4241	0.3606	0.2843
10	0.6139	0.5584	0.4632	0.3855	0.3220	0.2472
15	0.4810	0,4173	0.3152	0.2394	0.1827	0.1229
20	0.3769	0.3118	0.2145	0.1486	0.1037	0.0611
25	0.2953	0.2330	0.1460	0.0923	0.0588	0.0304

Table 3.4: Applying the discount factors to project 1

Year	*Project 1 cash flow (£)*	*Discount factor @ 10%*	*Discounted cash flow (£)*
0	– 100,000	1.0000	– 100,000
1	10,000	0.9091	9,091
2	10,000	0.8264	8,264
3	10,000	0.7513	7,513
4	20,000	0.6830	13,660
5	100,000	0.6209	62,090
Net Profit	£50,000		NPV: £618

It is interesting to note that the net present values for projects 1 and 3 are significantly different – even though they both yield the same net profit and both have the same return on investment. The difference in NPV reflects the fact that, with project 1, we must wait longer for the bulk of the income.

The main difficulty with NPV for deciding between projects is selecting an appropriate discount rate. Some organizations have a standard rate but, where this is not the case, then the discount rate should be chosen to reflect available interest rates (borrowing costs where the project must be funded from loans) plus some premium to reflect the fact that software projects are inherently more risky than lending money to a bank. The exact discount rate is normally less important than ensuring that the same discount rate is used for all projects being compared. However, it is important to check that the ranking of projects is not sensitive to small changes in the discount rate.

Alternatively, the discount rate can be thought of as a target rate of return. If, for example, we set a target rate of return of 15% we would reject any project that did not display a positive net present value using a 15% discount rate. Any project that displayed a positive NPV would be considered for selection – perhaps by using an additional set of criteria where candidate projects were competing for resources.

Table 3.5: Three estimated project cash flows

Year	*Project A (£)*	*Project B (£)*	*Project C (£)*
0	– 8,000	– 8,000	– 10,000
1	4,000	1,000	2,000
2	4,000	2,000	2,000
3	2,000	4,000	6,000
4	1,000	3,000	2,000
5	500	9,000	2,000
6	500	– 6,000	2,000
Net Profit	4,000	5,000	6,000

Internal rate of return

One disadvantage of NPV as a measure of profitability is that, although it may be used to compare projects, it might not be directly comparable with earnings from other investments or the costs of borrowing capital. Such costs are usually quoted as a percentage interest rate. The internal rate of return (IRR) attempts to provide a profitability measure as a percentage return that is directly comparable with interest rates. Thus, a project that showed an estimated IRR of 10% would be worthwhile if the capital could be borrowed for less than 10% or if the capital could not be invested elsewhere for a return greater than 10%.

The IRR is calculated as that percentage discount rate that would produce an NPV of zero. It is most easily calculated using a spreadsheet or other computer program that provides functions for calculating the IRR. Microsoft Excel and Lotus, for example, both provide IRR functions which, provided with an initial guess or seed value (which may be zero), will search for and return an IRR.

Manually, it must be calculated by trial-and-error or estimated using the technique illustrated in Fig. 3.3. This technique consists of guessing two values (one either side of the true value) and using the resulting NPVs (one of which must be positive and the other negative) to estimate the correct value. Note that this technique will provide only an approximate value but, in many cases that can be sufficient to dismiss a project that has a small IRR or indicate that it is worth making a more precise evaluation.

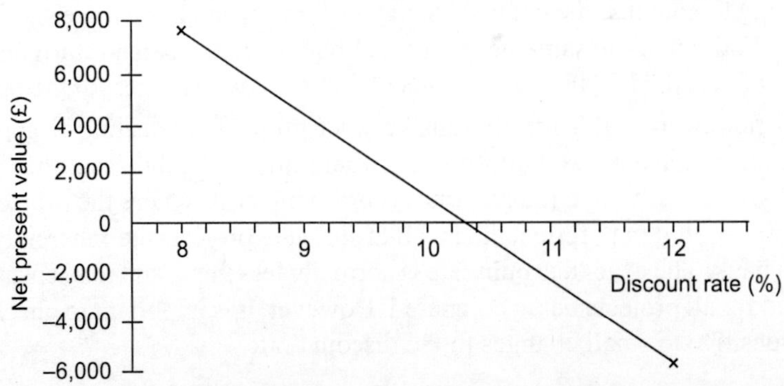

Fig. 3.3 Estimating the internal rate of return for project 1.

The internal rate of return is a convenient and useful measure of the value of a project in that it is a single percentage figure that may be directly compared with rates of return on other projects or interest rates quoted elsewhere.

Table 3.6 illustrates the way in which a project with an IRR of 10% may be directly compared with other interest rates. The cash flow for the project is shown in column (*a*). Columns (*b*) to (*e*) show that if we were to invest £100,000 now at an annual interest rate of 10% in, say, a bank, we could withdraw the same amounts as we would can, from the project at the end of each year, column (*e*), and we would be left with a net balance of zero at the end. In other words, investing in a project that has an IRR of 10% can produce exactly the same cash flow as lending the money to a bank at a 10% interest rate. We can therefore reason that a project with an IRR greater than current interest rates will provide a better rate of return than lending the investment to a bank. We can also say that it will be worth borrowing to finance the project if it has an IRR greater than the interest rate charged on the loan.

Table 3.6: A project cash flow treated as an investment at 10%

Year		*Equivalent investment at 10%*			
	(a) Project cash flow forecast (£)	*(b) Capital at start of year (£)*	*(c) Interest during year (£)*	*(d) Capital at end of year (£)*	*(e) End of year withdrawal (£)*
0	– 100,000	—	—	—	—
1	10,000	100,000	10,000	110,000	10,000
2	10,000	100,000	10,000	110,000	10,000
3	10,000	100,000	10,000	110,000	10,000
4	20,000	100,000	10,000	110,000	20,000
5	99,000	90,000	9,000	99,000	99,000
6		0	0	0	0

One deficiency of the IRR is that it does not indicate the absolute size of the return. A project with an NPV of £100,000 and an IRR of 15% can be more attractive than one with an NPV of £10,000 and an IRR of 18% – the return on capital is lower but the net benefits greater.

An often quoted objection to the internal rate of return is that, under certain conditions, it is possible to find more than one rate that will produce a zero NPV. This is not a valid objection since, if there are multiple solutions, it is always appropriate to take the lowest value and ignore the others. Spreadsheets will normally always return the lowest value if provided with zero as a seed value.

NPV and IRR are not, however, a complete answer to economic project evaluation.

- A total evaluation must also take into account the problems of funding the cash flows will we, for example, be able to repay the interest on any borrowed money and pay development staff salaries at the appropriate time?
- While a project's IRR might indicate a profitable project, future earnings from a project might be far less reliable than earnings from, say, investing with a bank. To take account of the risk inherent in investing in a project, we might require that a project earn a 'risk premium' (that is, it must earn, say, at least 15% more than current interest rates) or we might undertake a more detailed risk analysis as described in the following sections of this chapter.

3.7 RISK EVALUATION

***(MDU 2008-09, 2012-13)* (*10 Marks*)**

Every project involves risk of some form. When assessing and planning a project, we are concerned with the risk of the project's not meeting its objectives.

Risk identification and ranking

In any project evaluation we should attempt to identity the risks and quantify their potential effects. One common approach to risk analysis is to construct a project risk matrix utilizing a checklist of possible risks and to classify each risk according to its relative importance and likelihood. Note that the importance and likelihood need to be separately assessed–we might be less concerned with something that, although serious, is very unlikely to occur than with something less serious that is almost certain. Table 3.7 illustrates a basic project risk matrix listing some of the risks that might be considered for a project, with their importance and likelihood classified as high (H), medium (M), low (L) or exceedingly unlikely (—). So that projects may be compared the list of risks must be the same for each project being assessed. It is likely, in reality, that it would be somewhat longer than shown and more precisely defined.

The project risk matrix may be used as a way of evaluating projects (those with high risks being less favoured) or as a means of identifying and ranking the risks for a specific project.

Risk and net present value

Where a project is relatively risky it is common practice to use a higher discount rate to calculate net present value. This addition or risk premium, might, for example, be an additional 2% for a reasonably sale project or 5% for a fairly risks one. Projects, may be categorized as high, medium or low risk using a scoring method and risk premiums designated for each category. The premiums, even if arbitrary, provide a consistent method of taking risk into account.

Cost-benefit analysis

A rather more sophisticated approach to the evaluation of risk is to consider each possible outcome and estimate the probability of its occurring and the corresponding value of the outcome. Rather than a single cash flow forecast for a project, we will then have a set of cash flow forecasts, each with an associated probability of occurring. The value of the project is then obtained by summing the cost or benefit for each possible outcome weighted by its corresponding probability.

Table 3.7: A fragment of a basic project risk matrix

Risk	*Importance*	*Likelihood*
Software never completed or delivered	H	—
Project cancelled after design stage	H	—
Software delivered late	M	M
Development budget exceeded ≤ 20%	L	M
Development budget exceeded > 20%	M	L
Maintenance costs higher than estimated	L	L
Response time targets not met	L	H

This approach is frequently used in the evaluation of large projects such as the building of new motorways, where variables such as future traffic volumes, and hence the total benefit of shorter journey times, are subject to uncertainty. The technique does, of course, rely on our being able to assign probabilities of occurrence to each scenario and, without extensive study, this can be difficult.

When used to evaluate a single project, thc cost-benefit approach, by 'averaging out' the effects of the different scenarios, does not take account an organization's reluctance to risk damaging outcomes. Because of this, where overall profitability is the primary concern, it is often considered more appropriate for the evaluation of a portfolio of projects.

Risk profile analysis

An approach which attempts to overcome some of the objections to cost-benefit averaging is the construction of risk profiles using sensitivity analysis.

Table 3.8: BuyRight's income forecasts

Sales	*Annual sales income (£)* i	*Probability* p	*Excepted Value (£)* $i \times p$
High	800,000	0.1	80,000
Medium	650,000	0.6	390,000
Low	100,000	0.3	30,000
Expected Income			500,000

This involves varying each of the parameters that affect the project's cost or benefits to ascertain how sensitive the project's profitability is to each factor. We might, for example, vary one of our original estimates by plus or minus 5% and recalculate the expected costs and benefits for the project. By repeating this exercise for each of our estimates in turn we can evaluate the sensitivity of the project to each factor.

By studying the results of a sensitivity analysis we can identify those factors that are most important to the success of the project. We then need to decide whether we can exercise greater control over them or otherwise mitigate their effects. If neither is the case, then we must live with the risk or abandon the project.

Sensitivity analysis demands that we vary each factor one at a time. It does not easily allow us to consider the effects of combinations of circumstances, neither does it evaluate the chances of a particular outcome occurring. In order to do this we need to use a more sophisticated tool such as Monte Carlo simulation. There are a number of risk analysis applications available such as (*@Risk* from Palisade) that use Monte Carlo simulation and produce risk profiles of the type shown in Fig. 3.4.

Projects may be compared as in Fig. 3.4, which compares three projects with the same expected profitability. Project A is unlikely to depart far from this expected value compared to project B, which exhibits a larger variance. Both of these have symmetric profiles, which contrast with project C. Project C has a skewed distribution, which indicates that although it is unlikely ever to be much more profitable than expected, it is quite likely to be far worse.

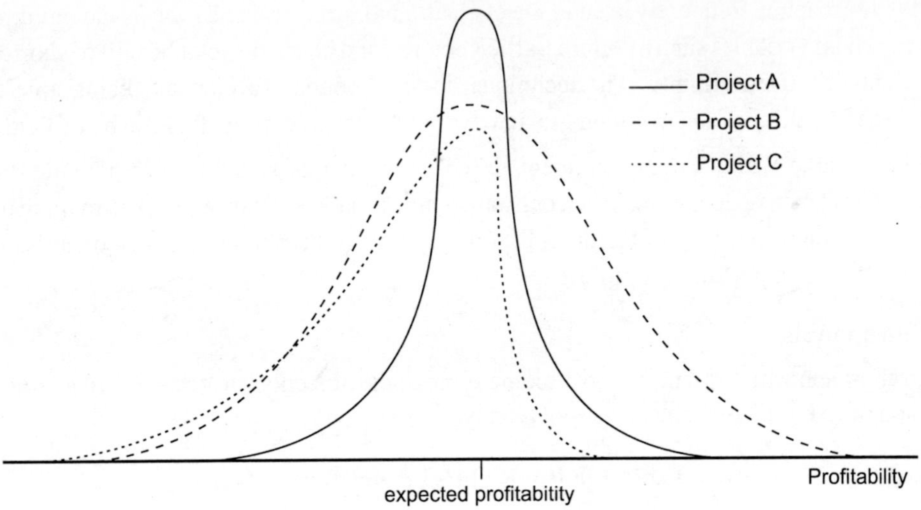

Fig. 3.4 A risk analysis profile.

Using decision trees

The approaches to risk analysis discussed previously rather assume that we are passive bystanders allowing nature to take its own course the best we can do is to reject over-risky projects or choose those with the best risk profile. There are many situations, however, where we can evaluate whether a risk is important and, if it is, indicate a suitable course of action.

Many such decisions will limit or affect future options and, at any point, it is important to be able to see into the future to assess how a decision will affect the future profitability of the project.

The decision largely rests upon the rate at which their equipment maintenance business expands – if their market share significantly increases (which they believe will happen if rumours of a competitor's imminent bankruptcy are fulfilled) the existing system might need to be replaced within 2 years. Not replacing the system in time could be an expensive option as it could lead to lost revenue if they cannot cope with the increase in invoicing demand. Replacing it immediately will, however, be expensive as it will mean deferring other projects that have already been scheduled.

They have calculated that extending the existing system will have an NPV of £57,000, although if the market expands significantly, this will be turned into a loss with an NPV of – £100,000 due to lost revenue. If the market does expand, replacing the system now has an NPV of £250,000 due to the benefits of being able to handle increased sales and other benefits such as improved management information. If sales do not increase, however, the benefits will be severely reduced and the project will suffer a loss with an NPV of – £50,000.

The company estimate the likelihood of the market increasing significantly at 20% – and, hence, the probability that it will not increase as 80%. This scenario can be represented as a tree structure as shown in Fig. 3.5.

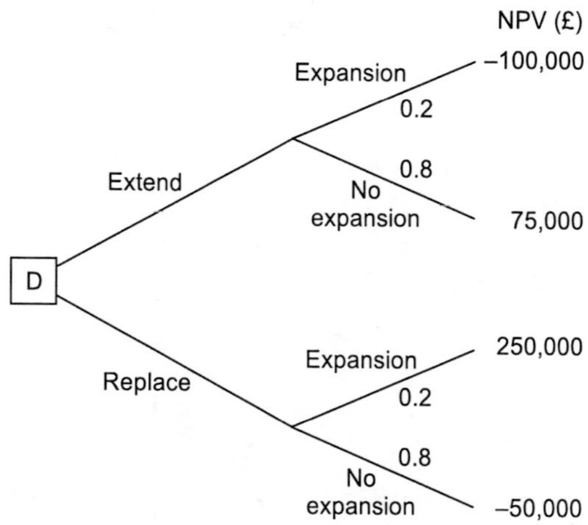

Fig. 3.5 A decision tree.

The analysis of a decision tree consists of evaluating the expected benefit of taking each path from a decision point (denoted by D in Fig. 3.5). The expected value of each path is the sum of the value of each possible outcome multiplied by its probability of occurrence. The expected value of extending the system is therefore £40,000 (75,000 × 0.8 – 100,000 × 0.2) and the expected value of replacing the system £10,000 (250,000 × 0.2 – 50,000 × 0.8).

This example illustrates the use of a decision tree to evaluate a simple decision at the start of a project. One of the great advantages of using decision trees to model and analyse problems is the ease with which they can be extended. Fig. 3.6 illustrates an extended version of decision tree, which includes the possibility of a later decision should they decide to extend the system and then find there is an early market expansion.

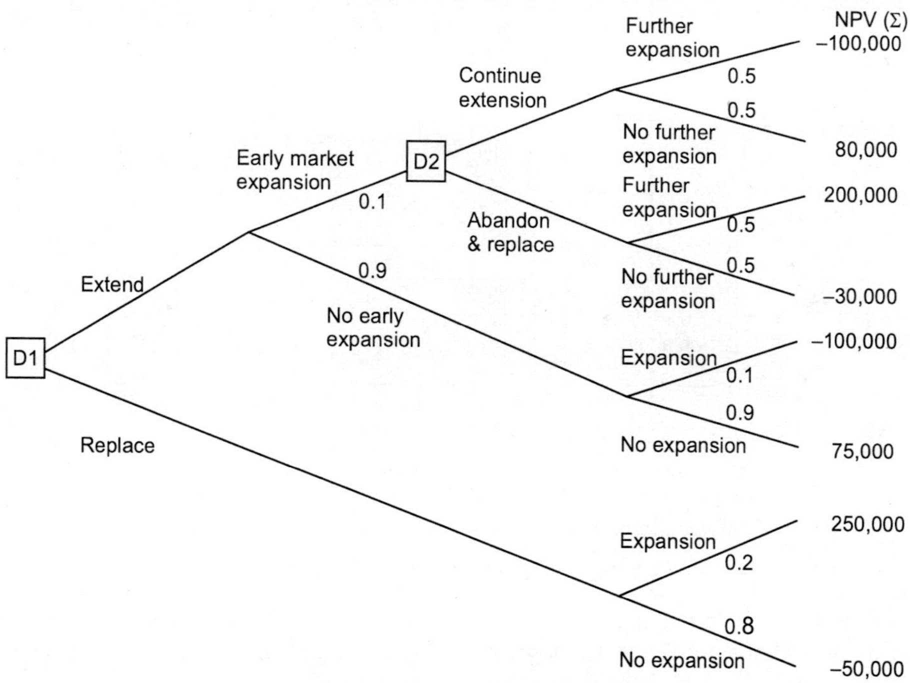

Fig. 3.6 An extension to decision tree.

3.8 SELECTION OF AN APPROPRIATE PROJECT REPORT

(MDU 2007-08, 2010-11, 2013-14) (08 Marks)

The development of software in-house suggests that the project has certain characteristics:

- the project team and the users belong to the same organization;
- the projects being considered slot into a portfolio of existing computer-based systems;
- the methodologies and technologies to be used are not selected by the project manager, but are dictated by local standards.

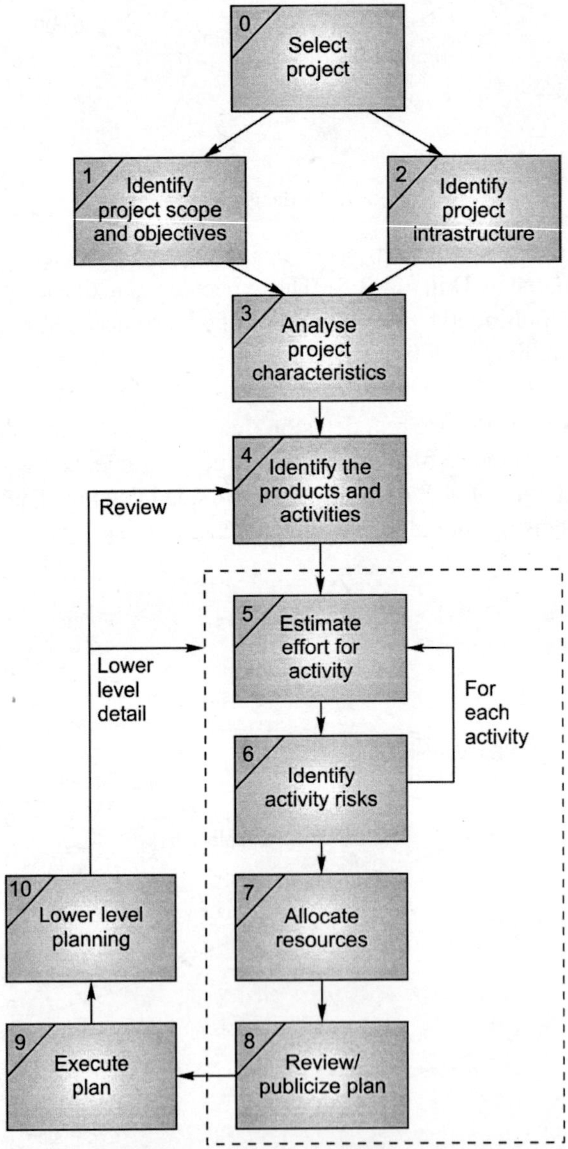

Fig. 3.7 Project analysis is the subject of Step 3.

However, where a series of development projects is being carried out by a software house for different external customers, the methodologies and technologies to be used will have to be decided for each individual project. This decision-making process has been called 'technical planning' by some, although here we use the term 'project analysis'. Even where development is in-house, it is important to spend some time looking for any characteristics of the new project that might make us take a different approach from that used on previous projects. It is this analysis that is the subject of this chapter.

The relevant part of the Step Wise approach is Step 3: Analyse protect characteristics. The selection of a particular process model will add new products to the Project Breakdown Structure or new activities to the activity network. This will create outputs for Step 4: Identify the products and activities of the project (see Fig. 3.7).

In the remainder of this chapter we will firstly look at how the characteristics of a project will influence the approach to the planning of a project. We will then look at some of the most common *process model*, namely the waterfall approach, prototyping and incremental delivery.

3.9 CHOOSING TECHNOLOGIES

An outcome of project analysis will be the selection of the most appropriate methodologies and technologies. Methodologies include techniques like the various flavours of object-oriented (OO) development. SSADM and JSP (Jackson Structured Programming) while technologies might include an appropriate application-building environment, or the use of knowledge-based system tools.

As well as the products and activities, the chosen technology will influence the following aspects of a project:

- the training requirements for development staff;
- the types of staff to be recruited;
- the development environment – both hardware and software;
- system maintenance arrangements.

We are now going to describe some of the steps of project analysis.

Identify project as either objectives-driven or product-driven *(Very Important)*

You will recall from Chapter 1 that we distinguished between objectives driven and product driven projects. Very often a product-driven project will have been preceded by an objectives-driven project which chose the general software solution that is to be implemented.

There will be cases where things are so vague that even the objectives of the project are uncertain or are the subject of disagreement. People may be experiencing a lot of problems but no-one knows exactly what the solution to the problems might be. It could be that the IT specialists can provide help in some places but assistance from other specialism is needed in others. In these kinds of situation a *soft systems* approach might need to be considered.

Analyse other project characteristics

The sorts of question that would need to be asked include the following:

- **Is a data orientated or a control orientated system to be implemented?** 'Data orientated' systems generally mean information systems that will have a considerable database. 'Control orientated' systems refer to embedded control systems. These days it is not uncommon to have systems with components of both types.
- **Will the software that is to be produced be a general package or application specific?** An example of a general package would be a spreadsheet or a word processing package. An application specific package could be, for example, an airline seat reservation system.

- **Is the system to be implemented of a particular type for which specific tools have been developed?** For example:
  - *does it involve concurrent processing?* – if so the use of techniques appropriate to the analysis and design of such systems would be considered;
  - *will the system to be created be knowledge based?* – expert systems have a set of rules which result in some 'expert advice' when applied to a problem domain (sets of methods and tools have been developed to assist in the creation of such systems); or
  - will the system to be produced make heavy use of computer graphics?
- **Is the system to be created safety-critical?** For instance, could a malfunction in the system endanger human life?
- **What is the nature of the hardware/software environment in which the system will operate?** It might be that the environment in which the final software will operate is different from that in which it will be developed. Embedded software may be developed on a large development machine that has lots of supporting software tools in the way of compilers, debuggers, static analysers and so on, but might then be down-loaded to a small processor in the larger engineered product. A system destined for a personal computer will need a different approach to one defined for a main-frame or a client-server environment.

Identify high level project risks *(Important)*

When we first embark on a project we might be expected to work out elaborate plans even though we are at least partially ignorant of many important factors that will affect the project. For example, until we do a detailed investigation of the users' requirements we will not be able to estimate how much effort will be needed to build a system to meet those requirements. The greater the uncertainties at the beginning of the project, the greater the risk that the project will be unsuccessful. Once we recognize a particular area of uncertainty we can, however, take steps to reduce its uncertainty.

One suggestion is that uncertainty can be associated with the *products*, *processes*, or *resources* associated with the project.

Product uncertainty. Here we ask how well the requirements are understood. It might be that the users themselves are uncertain about what a proposed information system is to do. The government, say, might introduce a new form of taxation but the way this is going to operate in detail will not be known until a certain amount of case law has been built up. Some environments can change so quickly that what was a precise and valid statement of requirements rapidly becomes out of date.

Process uncertainty. It might be that the project under consideration is the first where an organization has tried to use a method, such as SSADM or OMT, that is new to them. Perhaps a new application building tool is being used. Any change in the way that the systems are developed is going to introduce uncertainty.

Resource uncertainty. The main area of uncertainty here will almost surely be the availability of staff of the right ability and experience. A major influence on the degree of uncertainty in a project will be the sheer size of a project. The larger the number of resources needed or the longer the duration of the project, the more inherently risky it is likely to be.

Take into account user requirements concerning implementation

A user organization lays down standards that have to be adopted by any contractor providing software for them. For example the UK Civil Service favours the SSADM standard where information systems are being developed.

It is common for organizations to specify that suppliers of software have BS EN ISO 9001:1994 or TickIT accreditation. This will affect the way projects are conducted.

Select general life cycle approach

Control systems. A real-time system will have to be implemented using an appropriate methodology, for example, Mascot. Real-time systems that employ concurrent processing will use techniques such as Petri nets.

Information systems. Similarly, an information system will need a methodology, such as SSADM or Information Engineering, that matches that type of environment. SSADM will be especially appropriate where the project will employ a large number of development staff whose work will need to be coordinated: the method lays down in detail what needs to be done and what products need to be created at each step. Team members would therefore know exactly what is expected of them.

General applications. Where the software to be produced is for the general market rather than for a specific application and user, then a methodology such as SSADM would have to be thought about very carefully. This is because the trainers of the method make the assumption that a specific user exists. Some parts in the method also assume that there is an existing system that can be analysed to yield the logical features of the new, computer-based, system.

Specialized techniques. These have been invented to expedite the development of, for example, *knowledge-based systems* where there are a number of specialized tools and logic based programming languages that can be used to implement this type of system. Similarly a number of specialized techniques and tools have been developed to assist in the development of *graphics-based, systems.*

Hardware environment. The environment in which the system is to operate can put constraints on the way it is to be implemented. For instance, the need for a fast response time or for the software to take up only a small part of computer memory may mean that only low-level programming languages can be used—particularly in real-time and embedded systems.

Safety-critical systems. Where safety and reliability are of the essence, it might be possible to justify the additional expense of a formal specification using a notation such as Z or VDM. Really critical systems call for expensive measures such as having independent teams develop parallel systems with the same functionality. The parallel systems can then run concurrently when the application is in operation so that the results of each of the parallel systems can be cross-checked.

Imprecise requirements. Uncertainties or a *novel hardware/software platform* may mean that a *prototyping* approach should be considered. If the environment in which the system is to be implemented is a rapidly changing one, then serious consideration would need to be given to *incremental delivery.* If the users have *uncertain objectives* in connection with the project, then a *soft systems* approach might be desirable.

3.10 TECHNICAL PLAN CONTENTS LIST

(MDU 2009-10) (05 Marks)

The analysis described above will produce a number of practical requirements that will be fed into the next stage of the planning process. These requirements might add activities to the project and might involve the acquisition of items of software or hardware or the adoption of particular methodologies whichmight require staff training. As these recommendations imply certain costs, they should be recorded formally.

A preliminary version of this technical plan can be produced by a software house to help in the preparation of the bid for a contract. In some cases, it may actually be shown to the potential customer in order to show the basis for the bid price and generally to impress the customer with the soundness of the approach the software house intends to adopt.

The technical plan is likely to have the following contents:

1. Introduction and summary of constraints:
   (*a*) character of the system to be developed;
   (*b*) risks and uncertainties of the project;
   (*c*) user requirements concerning implementation.
2. Recommended approach:
   (*a*) selected methodology or process model;
   (*b*) development methods;
   (*c*) required software tools;
   (*d*) target hardware/software environment.
3. Implementation:
   (*a*) required development environment;
   (*b*) required maintenance environment;
   (*c*) required training.
4. Implications:
   (*a*) project products and activities – these will have an effect on the schedule duration and overall project effort;
   (*b*) financial – this report will be used to produce costings.

3.11 CHOICE OF PROCESS MODELS

The word 'process' is sometimes used to emphasize the idea of a system *in action.* In order to achieve an outcome, the system will have to execute one or more activities: this is its process. This idea can be applied to the development of computer-based systems where a number of interrelated activities have to be undertaken to create a final product. These activities can be organized in different ways and we can call these *process models.*

A major part of the planning will be the choosing of the development methods to be used and the slotting of these into an overall process model.

The planner needs not only to select methods but also to specify how the method is to be applied. With methods such as SSADM, there is a considerable degree of choice about how it is to be applied: not all parts of SSADM are compulsory. Many student projects have the rather basic failing that at the planning stage they claim that, say, SSADM is to be used: in the event, all that is produced are a few SSADM fragments such as a lop level data flow diagram and a preliminary logical data structure diagram. If this is all the particular project requires, it should be stated at the outset.

3.12 STRUCTURED METHODS

Although some 'object-oriented' specialists may object (!), we include the OO approach as a structured method – after all, we hope it is not unstructured. Structured methods are made up of sets of steps and rules, which, when applied produce system products such as data flow diagrams. Each of these products is carefully documented. Such methods are often time consuming compared to more intuitive approaches and this implies some additional cost. The pay-off is such things as a less error prone and more maintainable final system. This balance of costs and benefits is more likely to be justified on a large project involving many developers and users. This is not to say that smaller projects cannot justify the use of such methods.

3.13 RAPID APPLICATION DEVELOPMENT

(MDU 2008-09, 2010-11, 2012-13, 2013-14) *(10 Marks)*

The RAD (Rapid Application Development Model) model is proposed when requirements and solutions can be modularized as independent system or software components, each of which can be developed by different teams.

It might be thought that users would generally welcome the more professional approach that structured methods imply. However, customers of IS/IT are concerned with getting working business applications delivered quickly and at less cost and often see structured methods as unnecessarily bureaucratic and slow. A response to this has been *rapid application development* (RAD).

The RAD approach does not preclude the use of some elements of structured methods such as Logical Data Structure diagrams but also adopts tactics such as *joint application development* (JAD) workshops. In these workshops, developers and users work together intensively for, say, three to five days and identify and agree fully documented business requirements. Often, these workshops are conducted away from the normal business and development environments in *clean rooms,* that is, special conference rooms free from outside interruption and suitably furnished with white boards and other aids to communication. This is based on the belief that communication and negotiation that might take several weeks or months via a conventional approach of formal reports and proposals and counter-proposals can be speeded up in these hothouse conditions.

Another practice associated with RAD is *time-boxing* where the scope of a project is rigidly constrained by a pre-agreed and relatively close absolute deadline. Requirements that have to be omitted in order to meet this deadline are considered in later time-boxes.

Some of these ideas will be considered further in our later discussions on prototyping and the incremental approach.

3.14 THE WATERFALL MODEL

(MDU 2012-13) *(10 Marks)*

It is a very common software development process model. The waterfall model was popularized in the 1970s and permeates most current software engineering textbooks and standard industrial practices. It first appearance in the literature dates back to the late 1950s. As a result of experience gained in the development of a large air-defense software system called SAGE (Semi-Automated Ground Environment).

This is the 'classical' model of system development. An alternative name for this model is the *one-shot* approach. As can be seen from the example in Fig. 3.8, there is a sequence of activities working from top to bottom. The diagram shows some arrows pointing upwards and backwards. This indicates that a later stage might reveal the need for some extra work at an earlier stage, but this should definitely be the exception rather the rule. After all, the flow of a waterfall should be downwards with the possibility of just a little splashing back. The limited scope for iteration is in fact one of the strengths of this process model. With a large project you want to avoid having to go back and rework tasks that you thought had been completed. For a start, having to reopen what was previously thought to be a completed activity plays havoc with promised completion dates.

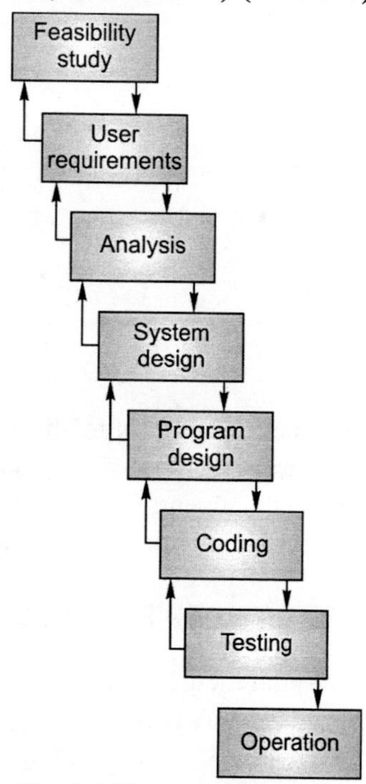

Fig. 3.8 The waterfall model.

We contend that there is nothing intrinsically wrong with the waterfall approach, even though more recent writers have suggested different models. It is the ideal that the project manager strives for. The waterfall approach allows project completion times to be forecast with more confidence than is the case with some more iterative approaches and this allows projects to be controlled effectively. However, where there is uncertainty about how a system is to be implemented, and unfortunately there very often is, a more flexible, iterative, approach may be required.

3.15 THE V-PROCESS MODEL *(Very Important)*

Fig. 3.9 gives a diagrammatic representation of this model. This is an elaboration of the waterfall model and stresses the necessity for validation activities that match the activities that create the products of the project.

The V-process model can be seen as expanding the testing activity in the waterfall model. Each step has a matching validation process that can, where defects are found, cause a loop back to the corresponding development stage and a reworking of the succeeding steps. Ideally, this feeding back should only occur where a discrepancy has been found between what was specified by a particular activity and what was actually implemented in the next lower activity on the descent of the V loop. For example, the system designer might have written that a calculation be carried out in a certain way. The person who structured the software that fulfilled this design might have misunderstood what was required.

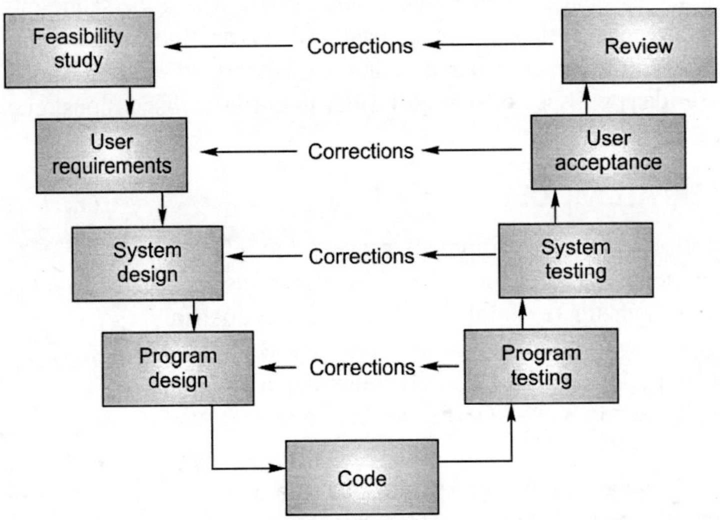

Fig. 3.9 The V-process model.

At system testing stage, the system designer would carry out checks that ensure that the software is doing what was specified in the design document and would discover the program designer's misreading of that document. Only corrections should be fed back, not the system designer's second thoughts, otherwise the project would be slipping into an "evolutionary prototyping" approach.

3.16 THE SPIRAL MODEL *(MDU 2008-09, 2011-12, 2012-13) (20 Marks)*

The spiral model was originally proposed by Boehm is an evolutionary software model that couples the iterative nature of prototyping with the controlled and systematic aspects of the linear segmental model.

It could be argued that this is another way of looking at the basic waterfall model. In the waterfall model, there is a possible escape at the end of any of the activities in the sequence. A feasibility study might decide that the implementation of a proposed system would be beneficial. The management therefore authorize work on the detailed collection and analysis of user requirements. Some analysis, for instance the interviewing of users, might already have taken place at the feasibility stage, but a more thorough investigation is now launched. This might reveal that in fact the costs of implementing the system would be higher than originally estimated and lead managers to decide to abandon the project.

A greater level of detail is considered at each stage of the project and a greater degree of confidence about the probability of success for the project should be justified. This can be portrayed as a loop or a spiral where the system to be implemented is considered in more and more detail in each sweep and an evaluation process is undertaken before the next iteration is embarked upon. Fig. 3.10 illustrates how SSADM can be interpreted in such a way.

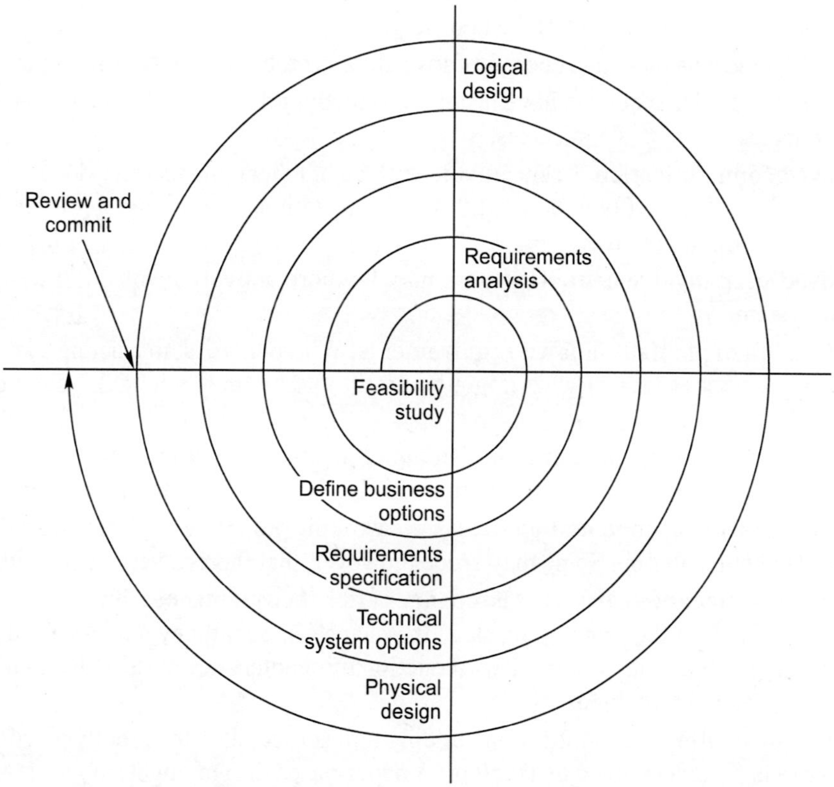

Fig. 3.10 The application of the spiral model to SSADM version 4.

The goal of the spiral model of the software production process is to provide a framework for designing such process, guided by the risk levels in the projects at hand. As opposed to the previously presented models, the spiral model may be viewed as a meta-model, because it can accomodate any process development model. By using it as a reference, one may choose the most appropriate development model (*e.g.*, evolutionary versus waterfall). The guiding principle behind such choice is the level of risk; accordingly, the spiral model provides a view of the production process that supports risk management.

3.17 SOFTWARE PROTOTYPING (*MDU 2008-09, 2011-12*) (*10 Marks*)

A prototype is a working model of one or more aspects of the projected system. It is constructed and tested quickly and inexpensively in order to test out assumptions.

Prototypes can be classified as throw-away, evolutionary or incremental.

Throw-away prototypes. Here the prototype is used only to test out some ideas and is then discarded when the development of the operational system is commenced. The prototype could be developed using a different software environment (for example, a 4 GL as opposed to a 3 GL for the final system where machine efficiency is important) or even on a different kind of hardware platform.

Evolutionary prototypes. The prototype is developed and modified until it is finally in a state where it can become the operational system. In this case, the standards that are used to develop the software have to be carefully considered.

Incremental prototypes. It could be argued that this is, strictly speaking, not prototyping. The operational system is developed and implemented in small stages so that the feedback from the earlier stages can influence the development of the later stages.

Some of the reasons that have been put forward for prototyping are the following:

Learning by doing. When we have just done something for the first time we can usually look back and see where we have made mistakes.

Improved communication. Users are often reluctant to read the massive documents produced by structured methods. Even if they do read this documentation, they do not get a feel for how the system is likely to work in practice.

Improved user involvement. The users may be more actively involved in design decisions about the new system.

Clarification of partially-known requirements. Where there is no existing system to mimic, users can often get a better idea of what might be useful to them in a potential system by trying out prototypes.

Demonstration of the consistency and completeness of a specification. Any mechanism that attempts to implement a specification on a computer is likely to uncover ambiguities and omissions.

Reduced need for documentation. Because a working prototype can be examined, there is less need for detailed documentation. Some may argue, however, that this is a very dangerous suggestion.

Reduced maintenance costs (that is, changes after the system goes live). If the user is unable to suggest modifications at the prototyping stage the chances are that they will ask for the changes as modifications to the operational system. This reduction of maintenance costs is the main plank in the financial case for creating prototypes.

Feature constraint. If an application building tool is used, then the prototype will tend to have features that are easily implemented by that tool. A paper-based design might suggest features that are expensive to implement.

Production of expected results. The problem with creating test runs is generally not the creation of the test data but the accurate calculation of the expected results. A prototype can be of assistance here.

Software prototyping is not without its drawbacks and dangers, however.

Users sometimes misunderstand the role of the prototype. For example, they might expect the prototype to be as robust as an operational system when incorrect data is entered or they might expect the prototype to have as fast a response as the operational system, although this was not the intention.

Lack of project standards possible. Evolutionary prototyping could just be an excuse for a sloppy 'hack it out and see what happens' approach.

Lack of control. It is sometimes difficult to control the prototyping cycle as the driving force could be the users' propensity to try out new things.

Additional expense. Building and exercising a prototype will incur additional expenses. The additional expense might be less than expected, however, because many analysis and design tasks would have to be undertaken anyway. Some research suggests that typically there is a 10% extra cost to produce a prototype.

Machine efficiency. A system built through prototyping, while sensitive to the users' needs, might not be as efficient in machine terms as one developed using more conventional methods.

Close proximity of developers. Prototyping often means that code developers have to be sited close to the users. One trend has been for organizations in developed countries to get program coding done cheaply in Third World countries such as India. Prototyping generally will not allow this.

3.18 OTHER WAYS OF CATEGORIZING PROTOTYPES *(MDU 2011-12, 2012-13) (10 Marks)*

What is being learnt?

The most important reason for having a prototype is that there is a need to learn about an area of uncertainty. For any prototype it is essential that the project managers define at the outset what it is intended to learn from the prototype.

This has a particular relevance to student projects. Students often realise that the software that they are to write as part of their project could not safely be used by real users. They therefore call the software a 'prototype'. However, if it is a real prototype then they must:

- specify what they hope to learn from the prototype;
- plan how the prototype is to be evaluated;
- report on what has actually been learnt.

Prototypes may be used to find out how a new development technique can be used. This would be the ease where a new methodology is being used in a pilot scheme. Alternatively, the development methods might be well-known, but the nature of the application might be uncertain.

Different projects will have uncertainties at different stages. Prototypes can therefore be used at different stages. A prototype might be used, for instance, at the requirements gathering stage to pin down requirements that seem blurred and shifting. A prototype might, on the other hand, be used at the design stage to test out the users' ability to navigate through a sequence of input screens.

To what extent is the prototyping to be done?

It would be unusual for the whole of the application to be prototyped. The prototyping might take MM of the following forms, which simulates only some aspects of the target application.

Mock-ups. For example, copies of the screens that the system is to use are shown to the users on a terminal, but the screens cannot actually be used.

Simulated interaction. For example, the user can type in a request to access a record and the system will respond by showing the details of a record, but the details shown are always the same and no access is made to a database.

Partial working model:

- *Vertical.* Some features are prototyped fully.
- *Horizontal.* All features are prototyped but not in detail (for example, there might not be a full validation of input).

What is being prototyped?

The human-computer interface. With information systems, what the system is to do has usually been established and agreed by management at a fairly early stage in the development of the system. Prototyping tends to be confined to establishing the precise way in which the operators are to interact with the system. In this case, it is important that the physical vehicle for the prototype be as similar as possible to the operational system.

The functionality of the system. In other cases, the precise way that the system should function internally will not be known. This will particularly be the case where a computer model of some real-world phenomenon is being developed. The algorithms used need to be repeatedly adjusted until they satisfactorily imitate the behaviour they should be modelling.

3.19 TOOLS

Special tools are not essential for prototyping but some have made it more practicable.

Application building tools have many features that allow simple computer based information systems to be set up quickly so that they can be demonstrated to the staff who will use them. An application written in a 3 GL becomes more and more complicated as changes accumulate in its code and this makes the software more difficult to alter safely, while applications implemented using application builders that are often form- and table-driven remain flexible.

It has been suggested that the ease of use of some 4GL application builders and the increasing IT awareness of end users might allow end users to create their own prototypes.

Where a prototype is being constantly tested by the users and modified by the developers there is a need to be able to control the different versions being produced and where necessary to back-track to previous versions.

3.20 A PROTOTYPING EXAMPLE *(Important)*

The use of prototyping on the COMET Commercial Estimating System project at the Royal Dockyards was the subject of a study, information about which has been published. The Royal Dockyards had been transferred to a system of 'commercial management' where they had to produce estimates of the cost of doing tasks that could then be compared against the cost of contracting the work out. Staff at the dockyards were inexperienced with this method of working.

It was decided to implement a computer system to support the estimating process and it was realised that the human-computer interface would be important. For this reason, the software house implementing the system suggested a prototyping approach.

Among the questions that had to be decided was who should be present at evaluation sessions. If managers were there, would they dominate the proceedings at the expense of those who would actually use the system? Furthermore, if the designers were present they might well feel defensive about the system presented and try and argue against changes suggested.

The results at the prototype

The impact of prototyping may be judged by the fact that although the preliminary design had been done using SSADM, as a result of the evaluation of the prototype the number of screens in the system was doubled.

A major problem was that of controlling changes to the prototype. In order to record and control the changes suggested by users, the changes were categorized into three types:

Cosmetic (about 35% of changes). These were simply changes to the layout of the screen. They were:

- implemented;
- recorded.

Local (about 60% of changes). These involved changes to the way that the screen was processed but did not affect other parts of the system. They were:

- implemented;
- recorded;
- backed-up so that they could removed at a later stage if necessary;
- inspected retrospectively.

Global (about 5% of changes). These were changes that affected more than one part of the processing. All changes here had to be the subject of a design review before they could be implemented.

3.21 INCREMENTAL DELIVERY

This is similar to the 'incremental prototyping' approach mentioned above. One of the most prominent advocates of this approach is Tom Gilb. The approach involves breaking the system down into small components which are then implemented and delivered in sequence. Each component that is delivered must actually give some benefit to the user. Fig. 3.11 gives a general idea of the approach.

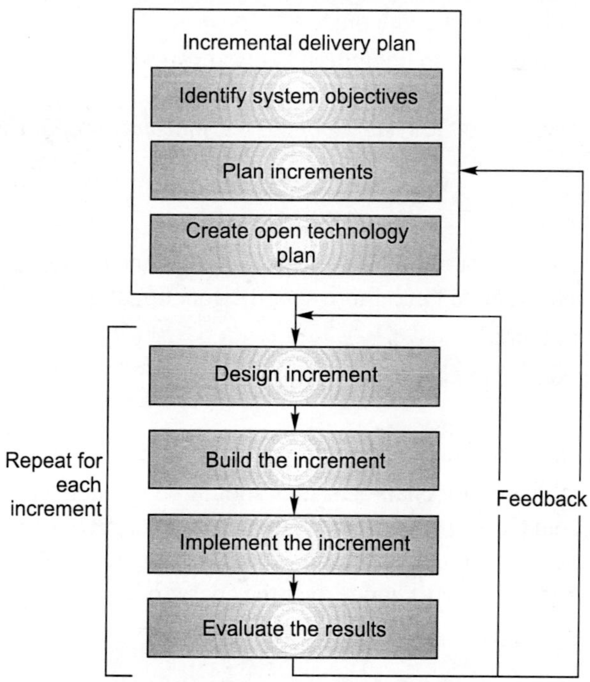

Fig. 3.11 Intentional incremental delivery.

Advantages of this approach

These are some of the justifications given for the approach:

- the feedback from early increments can influence the later stages;
- the possibility of changes in requirements is not so great as with large monolithic projects because of the shorter time-span between the design of a component and its delivery;
- users get benefits earlier than with a conventional approach;
- early delivery of some useful components improves cash flow, because you get some return on investment early on;
- smaller sub-projects are easier to control and manage;
- 'gold-plating', the requesting of features that are unnecessary and not in fact used, should be less as users will know that they get more than one opportunity to make their requirements known; if a feature is not in the current increment then it can be included in the next;
- the project can be temporarily abandoned if more urgent work crops up;
- job satisfaction is increased for developers who see their labours bearing fruit at regular, short, intervals.

Disadvantages

On the other hand these disadvantages have been put forward:

- 'software breakage', that is, later increments might require the earlier increments to be modified;
- developers might be more productive working on one large system than on a series of smaller ones.

The incremental delivery plan

The content of each increment and the order in which the increments are to be delivered to the users of the system have to be planned at the outset.

Basically the same process has to be undertaken as in strategic planning but at a more detailed level where the attention is given to increments of a user application rather than whole applications. The elements of the incremental plan are the *system objectives, incremental plan* and the *open technology plan.*

Identify system objectives

The purpose is to give an idea of the 'big picture', that is, the overall objectives that the system is to achieve. These can then expanded into more specific functional goals and quality goals.

Functional goals will include:

- objectives it is intended to achieve;
- jobs the system is to do;
- computer/non-computer functions to achieve them.

In addition, measurable quality characteristics should be defined, such as reliability, response and security. This reflects Tom Gilb's concern that system developers always keep sight of the objectives that they are trying to achieve on behalf of their clients. In the quickly changing environment of an application, individual requirements may change over the course of the project, but the objectives should not.

Plan increments

Having defined the overall objectives, the next stage is to plan the increments using the following guidelines:

- steps typically should consist of 1% to 5% of the total project;
- non-computer steps may be included – but these must deliver benefits directly to the users;
- ideally, an increment should take one month or less and should never take more than three months;
- each increment should deliver some benefit to the user;
- some increments will be physically dependent on others;
- value-to-cost ratios may be used to decide priorities (see table 3.9).

Very often a new system will be replacing an old computer system and the first increments may use parts of the old system, for example, the data for the database of the new system may initially be obtained from the old system's standing files.

Which steps should be first? Some steps might be prerequisites because of physical dependencies but others can be in any order. Value-to-cost ratios can be used to establish the order in which increments are to be developed. The customer is asked to rate each increment with a score in the range 1–10 in terms of its value. The developers also rate the cost of developing each of the increments with a score in the range 0–10. This might seem a rather crude way of evaluating costs and benefits, but people are often unwilling to be more precise. By then dividing the value rating by the cost rating, a rating which indicates the relative 'value for money' of each increment is derived.

The value to cost ratio = V/C where V is a score 1–10 representing value to customer and C is a score 0–10 representing cost.

Table 3.9: Ranking by value to cost ratio

Step	*Value*	*Cost*	*Ratio*	*Rank*
Profit reports	9	1	9	(2nd)
Online database	1	9	0.11	(6th)
Ad hoc enquiry	5	5	1	(4th)
Production sequence plans	2	8	0.25	(5th)
Purchasing profit factors	9	4	2.25	(3rd)
Clerical procedures	0	7	0	(7th)
Profit based pay for managers	9	0	∞	(1st)

Create open technology plan

If the system is to be able to cope with new components being continually added then it has to be built so that it is extendible, portable and maintainable.

As a minimum this will require the use of:

- a standard high level language;
- a standard operating system;
- small modules;
- variable parameters, for example, items such as organization name, department names and charge rates are held in a parameter file that can be amended without programmer intervention;
- a standard database management system.

These are all things that might be expected as a matter of course in a modern IS development environment.

3.22 AN INCREMENTAL EXAMPLE

Tom Gilb describes a project where a software house negotiated a fixed price contract with a three-month delivery time with the Swedish Government to supply a system to support map-making. It later became apparent that the original estimates of effort upon which the bid was probably about half what the real effort would be.

The project was replanned so that it was divided into ten increments, each supplying something of use to the customer. The final increments were not available until three months after the contract's delivery date. The customer was not in fact unhappy about this as the most important parts of the system had actually been delivered early.

3.23 SELECTING THE MOST APPROPRIATE PROCESS MODEL

Both Euromethod and ISO 12207 provide advice on this.

Euromethod distinguishes between the *construction* of an application and its *installation.* It is possible to use different approaches for these two stages. For example, an application could be constructed using a one-shot strategy but then be released to its users in increments. The only combinations of construction and installation strategies that are not feasible are evolutionary installation with any other construction approach than evolutionary.

In general, Euromethod suggests that where uncertainty is high then an evolutionary approach is to be favoured. An example of uncertainty would be where the users' requirements are not clearly defined. Where the requirements are relatively certain but there are many complexities, for example, where there is a large embedded system that is going to need a large amount of code, then an incremental approach might be favoured. Where deadlines are tight, then either an evolutionary or an incremental approach is favoured over a one-shot strategy, as both tactics should allow at least something to be delivered at the deadline, even if it is not all that was original promised.

3.24 SOFTWARE EFFORT ESTIMATION *(MDU 2008-09, 2011-12)* *(10 Marks)*

A software project estimate predicts the size of a new project, or the amount of effort required to complete a project. Software projects are usually delivered late because the project delivery times are usually under-estimated.

Software project estimation is the process of estimating various resources required for the completion of a project. Effective software project estimation is an important activity in any software development project. Under estimating software project and under staffing it often leads to low quality deliverables, and the project misses the target deadline leading to customer dissatisfaction and loss of credibility to the company. On the other hand, overstaffing a project without proper control will increase the cost of the project and reduce the competitveness of the company.

One definition of a successful project is that the system is delivered 'on time and within budget and with the required quality', which implies that targets are set and the project leader then tries to meet those targets. This assumes that the targets are reasonable—the possibility of a project leaders achieving record levels of productivity from the team, but still not meeting a deadline because of incorrect initial estimates is not recognized. Realistic estimates are therefore crucial to the project leader.

Estimating the effort required to implement software is notoriously difficult. Some of the difficulties of estimating are inherent in the very nature of software, especially its complexity and invisibility. In addition the intensely human activities that make up system development cannot be treated in a purely mechanistic way. Other difficulties include:

- **Novel applications of software.** With traditional engineering projects, it is often the case that the system to be created is similar to one constructed previously but for a different customer or in a different location. The estimates for such a project can therefore be based on previous experience. With software, in most major projects the product will in some way be unique and will therefore be clouded with doubts and uncertainties.
- **Changing technology.** For example, at the original maintenance billing system might have been written in Cobol, while the new extension for group accounts might be developed using an application building environment such as that provided by Oracle.
- **Lack of homogeneity of project experience.** As we will see, effective estimating should be based on information about how past projects have performed. It is surprising how many organizations do not make this data available to staff.

Table 3.10 is a set of figures recorded for actual projects carried out by the same organization.

Table 3.10: Some project data–effort in work months (as percentage of total effort in brackets)

Project	*Design*		*Coding*		*Testing*		*Total*	
	wm	*(%)*	*wm*	*(%)*	*wm*	*(%)*	*wm*	*SLOC*
a	3.9	(23)	5.3	(32)	7.4	(44)	16.7	6050
b	2.7	(12)	13.4	(59)	6.5	(26)	22.6	8363
c	3.5	(11)	26.8	(83)	1.9	(6)	32.2	13334
d	0.8	(21)	2.4	(62)	0.7	(18)	3.9	5942
e	1.8	(10)	7.7	(44)	7.8	(45)	17.3	3315
f	19.0	(28)	29.7	(44)	19.0	(28)	67.7	38988
g	2.1	(21)	7.4	(74)	0.5	(5)	10.1	38614
h	1.3	(7)	12.7	(66)	5.3	(27)	19.3	12762
i	8.5	(14)	22.7	(38)	28.2	(47)	59.5	26500

It would be very difficult on the basis of this information to advise a project manager about what sort of productivity to expect, or about the probable distribution of effort among the phases of design, coding and testing that could be expected from a new project.

There have been some attempts to set up industry-wide databases of past projects. However, this data seems to be of limited use to estimators as there are uncertainties in the way that various terms can be interpreted. For example, what exactly is meant by the term 'testing'? Does it cover the activity of the software developer when debugging code? Does 'design' include drawing up program structure diagrams or does this come under the heading of 'programming'?

Subjective nature of estimating. Some research shows that people tend to underestimate the difficulty of small tasks and overestimate that of large ones. In the world of software development this is perhaps justifiable, as large projects are usually disproportionately more complex and more difficult than smaller ones.

Political implications. Different groups within an organization have different objectives. The information systems development managers might, for example, want to see as many systems as possible implemented and will therefore put pressure on estimators to reduce cost estimates. One suggestion is that all estimates should be carried out within an organization by a specialist estimating group, independent of both the users and the project team. Not all agree with this, as staff involved in a project are more likely to be committed to targets where they have participated in formulating them.

3.25 WHERE ARE ESTIMATES DONE? *(MDU 2012-13) (10 Marks)*

Estimates are carried out at various stages of a software project. At each stage, the reasons for the estimate and the methods used will vary.

Strategic planning. The costs of computerizing potential applications as well as the benefits of doing so might need to be estimated to help decide what priority to give to each project. Such estimates might also influence the numbers of various types of development staff to be recruited by the organization.

Feasibility study. This ascertains that the benefits of the potential system will justify the costs.

System specification. Most system development methodologies usefully distinguish between the definition of the users' requirements and the design that documents how those requirements are to be fulfilled. The effort needed to implement different design proposals will need to be estimated. Estimates at the design stage will also confirm that the feasibility study is still valid, taking into account all that has been learnt during detailed requirements analysis.

Evaluation of suppliers' proposals. In the case of the maintenance group accounts subsystem, for example, might consider potting the actual system-building out to tender. Staff in the software houses that are considering a bid would need to scrutinize the system specification and produce estimates on which to base proposals.

Project planning. As the planning and implementation of the project progresses to greater levels of detail, more detailed estimates of smaller work components will be made. As well as confirming the earlier and more broad-brush estimates, these will help answer questions about, for example, when staff will have completed particular tasks and be available for new activities. Two general points can be made here:

- as the project proceeds, so the accuracy of the estimates should improve as knowledge about the nature of the project increases;
- conventional wisdom is that at the beginning of the project the user requirement (that is, a logical model of the required system) is of paramount importance and that premature consideration of the physical implementation is to be avoided. In order to do an estimate, however, the estimator will have to speculate about this physical implementation, for instance about the number of software modules to be written.

To set estimating into the context of the Step Wise framework (Fig. 3.12), re-estimating might take place almost any step, but specific provision is made for it at step 3. 'Analyse project characteristics', where a relatively high-level estimate will be produced, and in step 5 for each individual activity. As steps 5–8 are repeated at progressively lower levels, so estimates will be done at a finer degree.

3.26 PROBLEMS WITH OVER- AND UNDER-ESTIMATES

A project leader will need to be aware that the estimate itself, if known to the development team, will influence the time required to implement the system. An over-estimate might cause the project to take longer than it would otherwise. This can be explained by the application of two 'laws'.

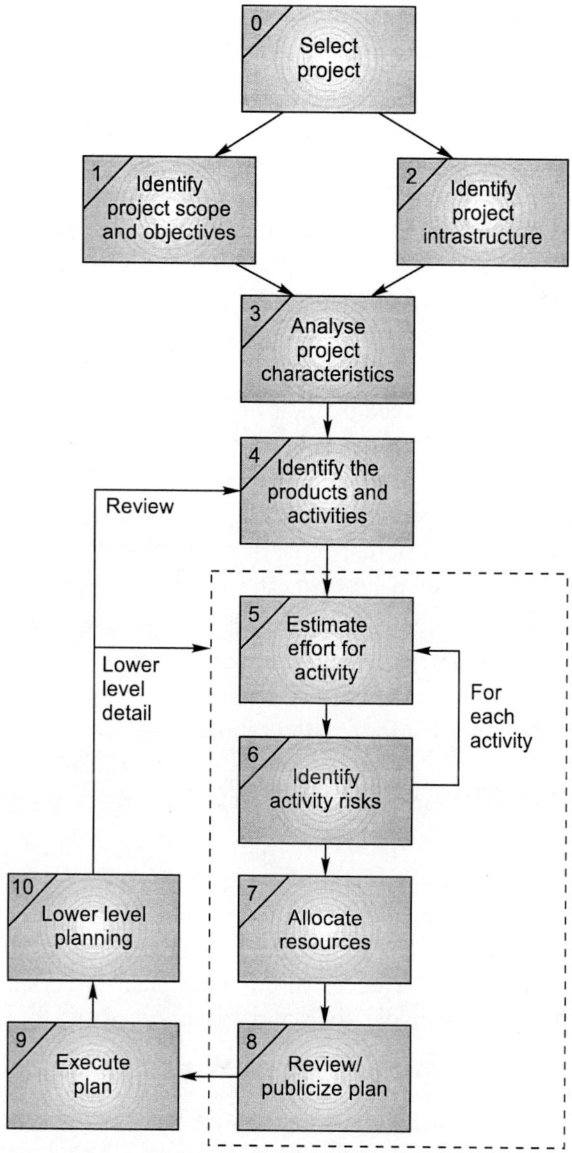

Fig. 3.12 Software estimation takes place in Steps 3 and 5 in particular.

Parkinson's Law. 'Work expands to fill the time available', which implies that given an easy target staff will work less hard.

Brooks' Law. The effort required to implement a project will go up disproportionately with the number of staff assigned to the project. As the project team grows in size so will the effort that has to go into management, coordination and communication. This has given rise, in extreme cases, to the notion of Brooks' Law. 'Putting more people on a late job makes it later'. If there is an over-estimate of the effort required then this might lead to more staff being allocated than are needed and managerial overheads will be increased. This is more likely to be of significance with large projects.

Some have suggested that while the under-estimated project might not be completed on time or to cost, it might still be implemented in a shorter time than a project with a more generous estimate. There must, however, be limits to this phenomenon where all the slack in the project is taken up.

The danger with the under-estimate is the effect on quality. Staff, particularly those with less experience, might respond to pressing deadlines by producing work which is sub-standard. Since we are into laws, this might be seen as a manifestation of Weinberg's zeroth law of reliability: 'if a system does not have to be reliable, it can meet any other objective'. In other words, if there is no need for the program actually to work, you can meet any programming deadline that might be set! Sub-standard work might only become visible at the later, testing, phases of a project, which are particularly difficult to control and where extensive rework can have catastrophic consequences for the project completion date.

Because of the possible effects on the behaviour of development staff caused by the size of estimates, they might be artificially reduced by their managers to increase pressure on staff. This will work only where staff are unaware that this has been done. Research has found that motivation and morale are enhanced where targets are achievable. If, over a period of time, staff become aware that the targets set are unattainable and that projects are routinely not meeting their published targets, then this will help to destroy motivation. Furthermore, people like to think of themselves as winners and there is a general tendency to put success down to our own efforts, while failure is blamed on the organization.

In the end, an estimate is not really a prediction, it is a *management goal.* Barry Boehm has suggested that if a software development cost is within 20% of the estimated cost estimate for the job then a good manager can turn it into a self-fulfilling prophecy. A project leader like Amanda will work hard to make the actual performance conform to the estimate.

3.27 THE BASIS FOR SOFTWARE ESTIMATING

The need for historical data

Nearly all estimating methods need information about how projects have been implemented in the past. However, care needs to be taken in judging the applicability of data to the estimators own circumstances because of possible differences in environmental factors such as the programming languages used, the software tools available, the standards enforced and the experience of the staff.

Measure of work

It is normally not possible to calculate directly the actual cost or time required to implement a project. The time taken to write a program might vary according to the competence or experience of the programmer. Implementation times might also vary because of environmental factors such as the software tools available. This usual practice is therefore to express the work content of the system to be implemented independently of effort, using a measure such as source lines of code (SLOC). The reader might also come across the abbreviation KLOC which refers to thousands of lines of code.

As can be imagined, SLOC is a very imprecise measure. Does it include comment lines? Are data declarations to be included? Unfortunately, researchers have not been consistent on these points. The writers' view is that comment lines are excluded, but data declarations are included. The argument for including data declarations is that in Cobol especially, it is possible to transfer much of the specification of processing to the Data Division, by using, for example, the report writer facility. Others might argue over this, but the main point is that consistency is essential.

Counting SLOC is difficult when employing application building tools, which often use tables or diagrams to record processing rules. Different measures of size are needed, such as object or function points, which are explained further on in this chapter.

Complexity

Two programs with the same KLOC will not necessarily take the same time to write, even if done by the same developer in the same environment. One program might be more complex. Because of this, SLOC estimates have to be modified to take complexity into account. Attempts have been made to find objective measures of complexity, but often it will depend on the subjective judgement of the estimator.

3.28 SOFTWARE EFFORT ESTIMATION TECHNIQUES *(Important)*

Barry Boehm, in his classic work on software effort models, identified the main ways of deriving estimates of software development effort as:

- **Algorithmic models.** Which use 'effort drivers' representing characteristics of the target system and the implementation environment to predict effort.
- **Expert judgement.** Where the advice of knowledgeable staff is solicited.
- **Analogy.** Where a similar, completed, project is identified and its actual effort is used as a basis for the new project.
- **Parkinson.** Which identifies the staff effort available to do a project and uses that as the 'estimate'.
- **Price to win.** Where the 'estimate' is a figure that appears to be sufficiently low to win a contract.
- **Top-down.** Where an overall estimate is formulated for the whole project and is then broken down into the effort required for component tasks.
- **Bottom-up.** Where component tasks are identified and sized and these individual estimates are aggregated.

Clearly, the 'Parkinson' method is not really an effort prediction method, but a method of setting the scope of a project. Similarly, 'price to win' is a way of deciding a price and not a prediction method. On these grounds, Boehm rejects them as prediction techniques although they might have some value as management techniques. There is, for example, a perfectly acceptable engineering practice of 'design to cost' which is one example of the broader approach of 'design by objectives'.

We will now look at some of these techniques more closely, first we will examine the difference between top-down and bottom-up estimating.

Bottom-up estimating

Estimating methods can be generally divided into bottom-up and top-down approaches. With the bottom-up approach, the estimator breaks the project into its component tasks and then estimates how much effort will be required to carry out each task. With a large project, the process of breaking down into tasks would be a repetitive one: each task would be analysed into its component sub-tasks and these in turn would be further analysed. This is repeated until you get to components that can be executed by a single person in about a week or two. The reader might wonder why this is not called a top-down approach: after all you are starting from the top and working down! Although this top-down analysis is an essential precursor to bottom-up estimating, it is really a separate one that of producing a Work Breakdown Structure (WBS). The bottom-up part comes in adding up the calculated effort for each activity to get an overall estimate.

The bottom-up approach is most appropriate at the later, more detailed, stages of project planning. If this method is used early on in the project cycle then the estimator will have to make some assumptions about the characteristics of the final system, for example the number and size of software modules. These will be working assumptions that imply no commitment when it comes to the actual design of the system.

Where a project is completely novel or there is no historical data available, the estimator would be well advised to use the bottom-up approach.

The top-down approach and parametric models

The top-down approach is normally associated with parametric (or algorithmic) models. These may be explained using the analogy of estimating the cost of rebuilding a house. This would be of practical concern to a house-owner who needs sufficient insurance cover to allow for rebuilding the property if it were destroyed. Unless the house-owner happens to be in the building trade it is unlikely that he or she would be able to work out how many brick layer-hours, how many carpenter-hours, electrician-hours and so on would be required. Insurance companies, however, produce convenient tables where the house-owner can find an estimate of rebuilding costs based on such *parameters* as the number of storeys and the floor space that a house has. This is a simple parametric model.

The effort needed to implement a project will be related mainly to variables associated with characteristics of the final system. The form of the parametric model will normally be one or more formulae in the form.

$$\text{effort} = (\text{system size}) \times (\text{productivity rate})$$

For example, system size might be in the form 'thousands of lines of code' (KLOC) and the productivity rate 40 days per KLOC. The values to be used will often be matters of subjective judgement.

A model to forecast software development effort therefore has two key components. The first is a method of assessing the size of the software development task to be undertaken. The second assesses the rate of work at which the task can be done.

Some parametric models, such as that implied by function points, are focused on system or task size, while others, such are COCOMO, are more concerned with productivity factors.

Having calculated the overall effort required, the problem is then to allocate proportions of that effort to the various activities within that project.

The top-down and bottom-up approaches are not mutually exclusive. Project managers will probably try to get a number of different estimates from different people using different methods. Some parts of an overall estimate could be derived using a lop-down approach while other parts could be calculated using a bottom-up method.

3.29 EXPERT JUDGEMENT

This is asking someone who is knowledgeable about either the application area or the development environment to give an estimate of the effort needed to carry out a task. This method will most likely be used when estimating the effort needed to change an existing piece of software. The estimator would have to earn out some kind of impact analysis in order to judge the proportion of code that would be affected and from that derive an estimate. Someone already familiar with the software would be in the best position to do this.

Some have suggested that expert judgement is simply a matter of guessing, but our own research has shown that experts tend to use a combination of an informal analogy approach where similar projects from the past are identified (see below), and bottom-up estimating.

3.30 ESTIMATING BY ANALOGY

The use of analogy is also called *case-based reasoning.* The estimator seeks out projects that have been completed (*sources cases*) and that have similar characteristics to the new project (the *target case*). The effort that has been recorded for the matching source case can then be used as a base estimate for the target. The estimator should then try to identify any differences between the target and the source and make adjustments to the base estimate for the new project.

This might be a good approach where you have information about some previous projects but not enough to draw generalized conclusions about what variables might make good size parameters.

A problem here is how you actually identify the similarities and differences between the different systems. Attempts have been made to automate this process. One software application that has been developed to do this is ANGEL. This identifies the source case that is nearest the target by measuring the Euclidean distance between cases. The source case that is at the shortest Euclidean distance from the target is deemed to be the closest match. The Euclidean distance is calculated:

$$\text{distance} = \text{square root of } ((\text{target-parameter}_1 - \text{source-parameter}_1)^2 + \ldots + (\text{target-parameter}_n - \text{source-parameter}_n)^2)$$

The above explanation is simply to give an idea of how Fuclidean distance might be calculated. The ANGEL package uses rather more sophisticated algorithms based on this principle.

3.31 ALBRECHT FUNCTION POINT ANALYSIS *(MDU 2010-11, 2012-13) (10 Marks)*

Function points and feature points are methods of estimating the "amount of functionality" required for a program, and are thus used to estimate project completion time. The basic idea involves counting inputs, outputs, and other features of a description of functionality.

This is a top-down method that was devised by Allan Albrecht when be worked for IBM. Albrecht was investigating programming productivity and needed some way to quantify the functional size of programs independently of the programming languages in which they had been coded. He developed the idea of *function points* (FPs).

The basis of function point analysis is that computer-based information systems comprise five major components, or *external user types* in Albrecht's terminology, that are of benefit to the users:

- **External input types** are input transactions that update internal computer files.
- **External output types** are transactions where data is output to the user. Typically these would be printed reports, since screen displays would come under external inquiry types (see below).
- **Logical internal file types** are the standing files used by the system. The term 'file' does not sit easily with modern information systems. It refers to a group of data that is usually accessed together. It might be made up of one or more *record types*. For example, a purchase order file might be made up of a record type PurchaseOrder plus a second that is repeated for each item ordered on the purchase order – PurchaseOrderItem.

- **External interface file types** allow for output and input that might pass to and from other computer applications, examples of this would be the transmission of accounting data from an order processing system to the main ledger system or the production of a file off direct debit details on a magnetic or electronic medium to be passed to the Bankers Automated Clearing System (BACS). Files shared among applications would also be counted here.
- **External inquiry types,** note the US spelling of inquiry, are transactions initiated by the user that provide information but do not update the internal files. The user inputs some information that directs the system to the details required.

The analyst has to identify each instance of each external user type in the projected system. Each component is then classified as having either high, average or low complexity. The counts of each external user type in each complexity band are multiplied by specified weights (see Table 3.11) to get FP scores, which are summed to obtain an overall FP count, which indicates the information processing size.

Table 3.11: Albrecht complexity multipliers

External user type	*Multiplier*		
	Low	*Average*	*High*
External input type	3	4	6
External output type	4	5	7
Logical internal file type	7	10	15
External interface file type	5	7	10
External inquiry type	3	4	6

One problem with FPs as originally defined by Albrecht was that the question of whether the external user type was of high, low or average complexity was rather subjective. The International FP User Group (IFPUG) have now promulgated rules on how this is to be judged. For example, in the case of logical internal files and external interface files, the boundaries shown in Table 3.12 are used to decide the complexity level.

Table 3.12: IFPUG file type complexity

Number of record types	*Number of data types*		
	< 20	*20 to 50*	*> 50*
1	low	low	average
2 to 5	low	average	high
> 5	average	high	high

Tables 3.13 and 3.14 are used to allocate scores to external inputs and external outputs. Each external inquiry has to be counted both as if it were an external input and an external output and whichever score is higher is used.

Table 3.13: IFPUG External input complexity

Number of file types accessed	*Number of data types accessed*		
	< 5	*5 to 15*	*> 15*
0 or 1	low	low	average
2	low	average	high
> 2	average	high	high

Table 3.14: IFPUG External output complexity

Number of file types	*Number of data types*		
	< 6	*6 to 19*	*> 19*
0 or 1	low	low	average
2 or 3	low	average	high
> 3	average	high	high

Function point analysis now goes on to take into account the fact that the effort required to implement a computer-based information system will relate not just to the number and complexity of the features to be provided but also to the environment in which the system is to operate.

Fourteen factors have been identified that can influence the degree of difficulty associated with implementing a system. The list that Albrecht produced related particularly to the concerns of information system developers in the late 1970s and early 1980s. Some technology that was then new and relatively threatening is now well established.

The technical complexity adjustment (TCA) calculation has had lots of problems. Some have even found that it produces less accurate estimates than using the unadjusted function point count. Because of these difficulties, we are going to omit further discussion of the TCA.

Tables have been calculated to convert the FPs to lines of code for various languages. For example, it is suggested that 91 lines of Cobol are needed on average to implement an FP while for C the figure is 128 and for QuickBasic is 64.

3.32 FUNCTION POINTS MARK II

(Important)

The Mark II method has been recommended by the CCTA (Central Computer and Telecommunications Agency), which lays down standards for UK government projects. At one time this Mark II approach seemed to be a good method to use with SSADM but some difficulties are now apparent. The 'Mark II' label implies an improvement and replacement of the Albrecht method. The Albrecht (now IFPUG) method, however, has had many refinements made to it and FPA Mark II remains a minority method used mainly in the UK.

As with Albrecht, the information processing size is initially measured in unadjusted function points (UFPs) to which a technical complexity adjustment can then be applied (TCA). The assumption here is that an information system comprises transactions that have the basic structure shown in Fig. 3.13.

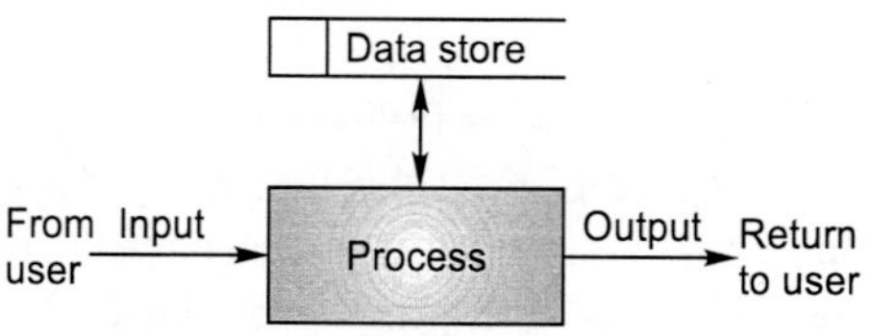

Fig. 3.13 Model of a transaction.

For each transaction the UFPs are calculated:

$W_i \times$ (number of input data element types) +
$W_e \times$ (number of entity types referenced) +
$W_o \times$ (number of output data element types)

Here, W_i, W_e, and W_o are weightings that might be derived by asking developers what proportion of effort has been spent in previous projects developing those parts of the software that deal with processing inputs, accessing and modifying stored data and processing outputs. From this it should be possible to work out the average hours of work generated by instances of each type of element.

The averages are then normalized into ratios, or weightings, which add up to 2.5. If this way of getting hold of the weightings seems too time-consuming then some industry averages are available, which are currently (1999) 0.58 for W_i, 1.66 for W_e, and 0.26 for W_o.

Mark II FPs follow the Albrecht method in recognizing that one system delivering the same functionality as another might be more difficult to implement (but also more valuable to the users) because of additional technical requirements, for example, the incorporation of additional security measures would increase the amount of effort to deliver the system. The original Albrecht FP method identified 14 technical complexity adjustment factors – Mark II FPs identify five more factors:

- interfaces to other applications;
- special security features;
- direct access for third parties;
- user training features;
- documentation requirements.

The addition of other factors to suit local circumstances is encouraged.

With both the Albrecht and Symons methods, FPs can be counted for previous projects where actual effort is known. If you have figures for the effort expended on past projects (in work-days for instance) and also the system sizes in FPs, you should be able to work out a productivity rate, that is:

productivity = size/effort

For new projects, the function points can be counted and then the effort can be projected using the productivity rate derived above:

effort = size/productivity

A more sophisticated way of doing this would be by using the statistical technique, least squares regression, to derive an equation in the form:

$$\text{effort} = \text{constant}_1 + \text{size} \times \text{constant}_2$$

Symons is very much against the idea of using function points to estimate SLOC rather than effort. One finding by Symons is that productivity, that is, the effort per function point to implement a system, is influenced very much by the size of the project. In general, larger projects, up to a certain point, tend to be more productive because of economies of scale. However, beyond a certain size they tend to become less productive because of additional management overheads.

Some of the rules and weightings used in FP counting, especially in the case of the Albrecht flavour, are rather arbitrary and have been criticized by academic writers on this account. FPs, however, are widely used in practice because of the lack of other methods of gauging the functional size of information systems.

3.33 OBJECT POINTS

This approach was devised at the Leonard N. Stern School of Business, New York University. It has similarities with the FP approach, but takes account of features that might be more readily identifiable if you are building a system using a high-level application building tool. The reader should be warned that despite its name, the technique has no direct bearing on object-oriented techniques. The approach uses counts of the screens, reports and 3GL components that an application might process– it is these that are referred to as *objects*. Each object has to be classified as one of the following:

- simple;
- medium;
- difficult.

Tables 3.15 and 3.16 show the scheme used to make this classification. The numbers of objects at each level are multiplied by the appropriate complexity weighting shown in Table 3.17. The weighted sub-totals are then summed to get an overall score for the application.

Table 3.15: Object points for screens

Number of views contained	*Number and source of data tables*		
	Total < 4 (< 2 servers < 3 clients)	*Total < 8 (< 3 servers 3 to 5 clients)*	*Total > 7 (> 3 servers > 5 clients)*
< 3	simple	simple	medium
3 to 7	simple	medium	difficult
> 7	medium	difficul	difficult

Table 3.16: Object Points for reports

Number of sections contained	*Number and source of data tables*		
	Total < 4 (< 2 servers < 3 clients)	*Total < 8 (< 3 servers 3 to 5 clients)*	*Total > 7 (> 3 servers > 5 clients)*
< 2	simple	simple	medium
2 or 3	simple	medium	difficult
> 3	medium	difficult	difficult

Table 3.17: Object Points complexity weightings

Object type	*Complexity weighting*		
	Simple	*Medium*	*Difficult*
Screen	1	2	3
Report	2	5	8
3GL components	—	—	10

Some of these objects might not need to be developed as there are already existing components that can be utilized The object point score can be adjusted to take this into account. Say that in an application containing 840 object points, 20% can be supplied by using existing components, then the adjusted new object points (NOP) score would be:

$$NOP = 840 \times (100 - 20)/100 = 672$$

Finally a productivity rate (PROD) has to be identified. It would be best if the estimator could use details of past projects to derive this. As an example, the developers of object points have published the details in Table 3.18 to calculate PROD. In the situation where this information was gathered, as the Case tool's features were improving with successive releases, so the experience of the developers with the tool was growing too.

Table 3.18: Object point effort conversion

Developer's experience and Capability/ICASE maturity and capability	*Very low*	*Low*	*Nominal*	*High*	*Very high*
PROD	4	7	13	25	50

An estimate of the person-months needed to carry out the project is then calculated by dividing PROD into NOP. For example, given the 672 new object points above and a development environment where productivity was nominal, then the estimated effort for the project would be 672/13 = 52 months.

3.34 A PROCEDURAL CODE-ORIENTED APPROACH

The previous approach would be useful at the design stage of a project and where a procedural programming language is not the primary vehicle for development. However how could you estimate the effort to develop an individual software module using a procedural language? An approach might be based on the following steps:

1. Envisage the number and type of programs in the final system. This is easiest where the system is of a conventional and well understood nature. Most information systems are built from a small set of system operations, such as insert, amend, update, display, delete, print. The same principle should equally apply to embedded systems, albeit with a different set of primitive functions.

2. Estimate the SLOC of each identified program. The estimator must have a particular implementation language in mind for this step.

One way to judge the number of instructions likely to be in a program is to draw up a program structure diagram and to visualize how many instructions would be needed to implement each identified procedure. The estimator might look at existing programs that have a similar functional description to help in this process.

Where programs for an information system are similar (for instance, they are data validation programs) then the number of data item types processed by each program is likely to be the major influence on size.

3. Estimate the work content, taking into account complexity and technical difficulty. The practice is to multiply the SLOC estimate by a factor for complexity and technical difficulty. This factor will depend largely on the subjective judgement of the estimator, for example, the requirement to meet particular highly constrained performance targets can greatly increase programming effort.

A weighting can be given when there is uncertainty, for example, about a new technique used in particular module, but this should not be excessive. Where there is a large amount of uncertainty then specific measures should be taken to reduce this by such means as the use of exploratory prototypes.

4. Calculate the work-days effort. Historical data can be used to provide ratios to convert weighted SLOC to effort. These conversion factors are often based on the productivity of a 'standard programmer' of about 15–18 months of experience. In installations where the rate of turnover is lower and the average programmer experience is higher this might be reflected in the conversion rate employed.

Note that the steps above can be used to derive an estimate of lines of code that can be used as an input to onc of the COCOMO models, which are now about to be described.

3.35 COCOMO: A PARAMETRIC MODEL *(MDU 2007-08, 2011-12) (10 Marks)*

Boehm's COCOMO (COnstructive COst MOdel) is often referred to in the literature on software project management, particularly in connection with software estimating. The term COCOMO really refers to a group of models.

Boehm originally based his models in the late 1970s on a study of 63 projects. Of these only seven were business systems and so they could be used with applications other than information systems. The basic model was built around the equation

$$\text{effort} = c \times \text{size}^k$$

where effort is measured in *pm*, or the number of 'person-months' consisting of units of 152 working hours, size is measured in *kdsi*, thousands of delivered source code instructions, and *c* and *k* are constants.

The first step was to derive an estimate of the system size in terms of *kdsi*. The constants, *c* and *k* (see Table 3.19), depended on whether the system could be classified, in Boehm's terms, as 'organic', 'semi-detached' or 'embedded'. These related to the technical nature of the system and the development environment.

- **Organic mode:** this would typically be the case when relatively small teams developed software in a highly familiar in-house environment and when the system being developed was small and the interface requirements were flexible.
- **Embedded mode:** this meant the product being developed had to operate within very light constraints and changes to the system were very costly.
- **Semi-detached mode:** this combined elements of the organic and the embedded modes or had characteristics that came between the two.

Table 3.19: COCOMO constants

System type	*c*	*k*
Organic	2.4	1.05
Semi-detached	3.0	1.12
Embedded	3.6	1.20

The exponent value *k,* when it is greater than 1, means that larger projects are seen as requiring disproportionately more effort than smaller ones. This reflected Boehm's finding that larger projects tended to be less productive than smaller ones because they needed more effort for management and coordination.

Boehm in fact found this, by itself, to be a poor predictor of the effort required and so went on to develop the intermediate version of COCOMO, which took into account 15 cost drivers. In the intermediate model, a nominal effort estimate, (pm_{nom}) is derived in a similar way as for the basic model.

The nominal estimate is then adjusted by a development effort multiplier (*dem*):

$$pm_{est} = pm_{nom} \times dem$$

where *dem* is calculated by taking into account multipliers based on the effort drivers in Table 3.20.

Table 3.20: COCOMO81 intermediate cost drivers

Driver type	*Code*	*Cost driver*
Product attributes	RELY	required software reliability
	DATA	database size
	CPLX	product complexity
Computer attributes	TIME	execution lime constraints
	STOR	main storage constraints
	VIRT	virtual machine volatility—degree to which the operating system changes
	TURN	computer turn around time
Personnel attributes	ACAP	analyst capability
	AEXP	application experience
	PCAP	programmer capability
	VEXP	virtual machine (*i.e.*, operating system) experience
	LEXP	programming language experience
Project attributes	MODP	use of modern programming practices
	TOOL	use of software tools
	SCED	required development schedule

These multipliers take into account such influences on productivity as Boehm's suggestion that having a programming team fully conversant with the programming language to be used could reduce the effort required to implement the project by up to 20% compared to a team with a very low or initially non-existent familiarity with the programming language. In fact, the biggest influence on productivity according to Boehm is the capability of the implementation team.

As an example of the approach, an organization might decide to use the following multipliers for assessing the effect of analyst capability (ACAP).

Very low	1.46
Low	1.19
Nominal	1.00
High	0.80
Very High	0.71

If the analysis involved in a project, taken as a whole, generally possess above average talent and productivity then the estimator might rate the ACAP as high and use a multiplier of 0.8, effectively reducing the nominal estimate by 20%.

The overall *dem* is calculated by multiplying the multipliers selected for each cost driver in Table 3.20 to create a single combined multiplier.

A new family of models, COCOMO II, is currently (1999) being refined by Barry Boehm and his co-workers. This approach uses various multipliers and exponents, the values of which have been set initially by experts. However, a database containing the performance details of executed projects is being built up and periodically analysed so that the expert judgements can be progressively replaced by values derived from actual projects. The new models take into account that there is now a wider range of process models in common use for software development projects than in the late 1970s and early 1980s. As we noted earlier, estimates are required at different stages in the system life cycle and COCOMO II has been designed to accommodate this by having models for three different stages.

Application composition. Where the external features of the system that the users will experience are designed. Prototyping will typically be employed to do this. With small applications that can be built using high-productivity application-building tools, development can stop at this point.

Early design. Where the fundamental software structures are designed. With larger, more demanding systems, where, for example, there will be large volumes of transactions and performance is important, careful attention will need to be paid to the architecture to be adopted.

Post architecture. Where the software structures undergo final construction, modification and tuning to create a system that will perform as required.

To estimate the effort for *application composition,* the counting of object points, which were described earlier, is recommended by the developers of COCOMO II.

At the *early design* stage, FPs are recommended as the way of gauging a basic system size. An FP count might be converted to a SLOC equivalent by multiplying the FPs by a factor for the programming language that is to be used.

The following model can then be used to calculate an estimate of person-months.

$$pm = \text{A} \times \text{size}^{sf} \times em_1 \times em_2 \times \ldots \times em_n$$

where *pm* is the effort in 'person-months', A is a constant (in 1998 it was 2.45), size is measured in SLOC (which might have been derived from an FP count as explained above), and *sf* is exponent scale factor.

The scale factor is derived thus:

$$sf = 1.01 + 0.01 \times \Sigma(\text{exponent driver ratings})$$

The qualities that govern the exponent drivers used to calculate the scale factor are listed below. Note that the less each quality is applicable, the bigger the value given to the exponent driver. The fact that these factors are used to calculate an exponent implies that the *lack* of these qualities will *increase* the effort required disproportionately more on larger projects.

- **Precedentedness.** This quality refers to the degree to which there are *precedents,* similar cases in the past, for the project that is being planned. The greater the novelty of the new system, the more uncertainty there is and the higher the value given to the exponent driver.
- **Development flexibility.** This is the degree to which the requirements can be met in a number of different ways. The less flexibility there is, the higher the value of the exponent driver.
- **Architecture/risk resolution.** This relates to the degree of uncertainty there is about the requirements. If they are not firmly fixed and are liable to change then this would lead to a high value being given to this exponent driver.
- **Team cohesion.** This reflects the degree to which there is a large dispersed team (perhaps in several countries) as opposed to there being a small tightly knit team.
- **Process maturity.** The chapter on software quality explains the process maturity model. The more structured and organized the way the software is produced, the lower uncertainty and the lower the rating will be for this exponent driver.

In the COCOMO II model the *effort multipliers, em*, are similar in nature to the development effort multipliers, *dem,* used in the original COCOMO. There are seven of these multipliers that are relevant to early design and sixteen that can be used at the post architecture stage. Table 3.21 lists the effort multipliers for early design and Table 3.22 for post architecture. As with COCOMO 81, each of these multipliers might, for a particular application, be given a rating of very low, low, nominal, high or very high. Each rating for each effort multiplier has a value associated with it. A value greater than 1 means that development effort is increased, while a value less than 1 causes effort to be decreased. The nominal rating means that the multiplier has no effect on the estimate, that is, it is 1. The intention is that the ratings that these and other values use in COCOMO II will be modified and refined over time as details of actual projects are added to the database.

Table 3.21: COCOMOII Early Design effort multipliers

Code	*Effort modifier*
RCPX	Product reliability and complexity
RUSE	Required reusability
PDIF	Platform difficulty
PERS	Personnel capability
PREX	Personnel experience
FCIL	Facilities available
SCED	Schedule pressure

Table 3.22: COCOMOII Post Architecture effort multipliers

Modifier type	*Code*	*Effort modifier*
Product attributes	RELY	Required software reliability
	DATA	Database size
	DOCU	Documentation match to life-cycle needs
	CPLX	Product complexity
	REUSE	Required reusability
Platform attributes	TIME	Execution time constraint
	STOR	Main storage constraint
	PVOL	Platform volatility
Personnel attributes	ACAP	Analyst capabilities
	AEXP	Application experience
	PCAP	Programmer capabilities
	PEXP	Platform experience
	LEXP	Programming language experience
	PCON	Personnel continuity
Project attributes	TOOL	Use of software tools
	SITE	Multisite development

CONCEPTUAL SHORT QUESTIONS WITH ANSWERS

1. CBA stands for?

Ans. CBA stands for Cost Benefit Analysis.

2. Compute the FP value for a project with the following information domain characteristics

No. of user inputs = 32

No. of user outputs = 60

No. of user inquiries = 24

No. of user files = 08

No. of external interfaces = 2

Assume that all complexity adjustment values are average. Assume 14 algorithms have been counted. Compute the feature point under same conditions.

Ans.

S. No.	*Domain characteristics*	*Count*	×	*Weighting factor*	
1.	No. of user inputs	32	×	4	128
2.	No. of user outputs	60	×	5	300
3.	No. of user inquiry	24	×	4	96
4.	No. of files	8	×	10	80
5.	No. of external interface	2	×	7	14

As complexity adjustment values are average $\sum_{i=1}^{14} Fi = 14 \times 3 = 42$.

$$\therefore \quad \text{FP} = \text{Count Total} \times [0.65 + 0.01 \times \Sigma Fi]$$
$$= 618 \times (0.65 + 0.01 \times 42) = 661.26.$$

3. What is pricing to win approach?

Ans. In pricing to win approach, the cost of the project is proportional to the budget of the customers. So, the estimated effort does not depend upon the software functionality but also on customer's capability to spend. This method results in cost over runs, time slippages and quality software.

4. Project Management involves four major activities. What are they?

Ans. (*a*) Feasibility Study (*b*) Project Planning

(*c*) Project Execution (*d*) Project Termination.

REVIEW QUESTIONS

1. What is the critical difference between the waterfall and spiral model? In the following cases, which performance model would you choose:

(*a*) Design a sales forecasting model and software to run this model.

(*b*) Insurance chain settlement process management.

2. Write short notes on:

(*a*) Spiral Model

(*b*) FP Analysis

(*c*) RAD

3. Describe FP analysis.
4. Write short notes on following:
   (*a*) Software prototyping
   (*b*) Spiral model for software development.
5. Explain Albrecht function point analysis for finding the effort required to complete a Software Project. What are the problems associated with over and under estimation.
6. What is pricing to win approach?
7. What is ESTIMATICS?
8. List down the steps that will be followed for network-based project management methodology.
9. If you were asked as an expert to provide an estimate to the effort needed to make certain change to an existing piece of software, what information would you like to have to hand to assist you in making that estimate?
10. What, is broad outline, would be the most suitable approach for each of the following?
    (*a*) a system which calculate the amount of a drug that should be administerted to a patient who has a particular complaint;
    (*b*) a system to administer a student loans scheme;
    (*c*) a system to control trains in the channel tunnel.

CHAPTER 4

ACTIVITY PLANNING

INSIDE THIS CHAPTER

4.1 INTRODUCTION

In earlier chapters we looked at methods for forecasting the effort required for a project – both for the project as a whole and for individual activities. A detailed plan for the project, however, must also include a schedule indicating the start and completion times for each activity. This will enable us to:

- ensure that the appropriate resources will be available precisely when required;
- avoid different activities competing for the same resources at the same time;
- produce a detailed schedule showing which staff carry out each activity;
- produce a detailed plan against which actual achievement may be measured;
- produce a timed cash flow forecast;
- replan the project during its life to correct drift from the target.

To be effective, a plan must be stated as a set of targets, the achievement or non-achievement of which can be unambiguously measured. The activity plan does this by providing a target start and completion date for each activity (or a window within which each activity may be carried out). The starts and completions of activities must be clearly visible and this is one of the reasons why it is advisable to ensure that each and every project activity produces some tangible product or 'deliverable'. Monitoring the project's progress is then, at least in part, a case of ensuring that the products of each activity are delivered on time.

As a project progresses it is unlikely that everything will go according to plan. Much of the job of project management concerns recognizing when something has gone wrong, identifying its causes and revising the plan to mitigate its effects. The activity plan should provide a means of evaluating the consequences of not meeting any of the activity target dates and guidance as to how the plan might most effectively be modified to bring the project back to target. We shall see that the activity plan may well also offer guidance as to which components of a project should be most closely monitored.

4.2 THE OBJECTIVES OF ACTIVITY PLANNING *(MDU 2008-09, 2011-12) (10 Marks)*

In addition to providing project and resource schedules, activity planning aims to achieve a number of other objectives which may be summarized as follows:

- **Feasibility assessment.** Is the project possible within required timescales and resource constraints? It is not until we have constructed a detailed plan that we can forecast a completion date with any reasonable knowledge of its achievability. The fact that a project may have been estimated as requiring two work-years effort might not mean that it would be feasible to complete it within, say, three months were eight people to work on it – that will depend upon the availability of staff and the degree to which activities may be undertaken in parallel.
- **Resource allocation.** What are the most effective ways of allocating resources to the project and when should they be available? The project plan allows us to investigate the relationship between timescales and resource availability (in general, allocating additional resources to a project shortens its duration) and the efficacy of additional spending on resource procurement.
- **Detailed costing.** How much will the project cost and when is that expenditure likely to take place? After producing an activity plan and allocating specific resources, we can obtain more detailed estimates of costs and their timing.
- **Motivation.** Providing targets and being seen to monitor achievement against targets is an effective way of motivating staff, particularly where they have been involved in setting those targets in the first place.
- **Co-ordination.** When do the staff in different departments need to be available to work on a particular project and when do staff need to be transferred between projects? The project plan, particularly with large projects involving more than a single project team, provides an effective vehicle for communication and co-ordination among teams. In situations where staff may need to be transferred between project teams (or work concurrently on more than one project), a set of integrated project schedules should ensure that such staff are available when required and do not suffer periods of enforced idleness.

Activity planning and scheduling techniques place an emphasis on completing the project in a minimum time at an acceptable cost or, alternatively, meeting an arbitrarily set target date at minimum cost. These are not, in themselves, concerned with meeting quality targets, which generally impose constraints on the scheduling process.

One effective way of shortening project durations is to carry out activities in parallel. Clearly we cannot undertake all the activities at the same time – some require the completion of others before they can start and there are likely to be resource constraints limiting how much may be done simultaneously. Activity scheduling will, however, give us an indication of the cost of these constraints in terms of lengthening timescales and provide us with an indication of how timescales may be shortened by relaxing those constraints. It is up to us, if we try relaxing precedence constraints by, for example, allowing a program online task to commence before the design has been completed, to ensure that we are clear about the potential effects on product quality.

4.3 WHEN TO PLAN

Planning is an ongoing process of refinement, each iteration becoming more detailed and more accurate than the last. Over successive iterations, the emphasis and purpose of planning will shift.

During the feasibility study and project start-up, the main purpose of planning will be to estimate timescales and the risks of not achieving target completion dates or keeping within budget. As the project proceeds beyond the feasibility study, the emphasis will be placed upon the production of activity plans for ensuring resource availability and cash flow control.

Throughout the project, until the final deliverable has reached the customer, monitoring and replanning must continue to correct any drift that might prevent meeting time or cost targets.

4.4 PROJECT SCHEDULES

(MDU 2012-13) (10 Marks)

Before work commences on a project or, possibly, a stage of a larger project, the project plan must be developed to the level of showing dates when each activity should start and finish and when and how much of each resource will be required. Once the plan has been refined to this level of detail we call it a *project schedule*. Creating a project schedule comprises four main stages.

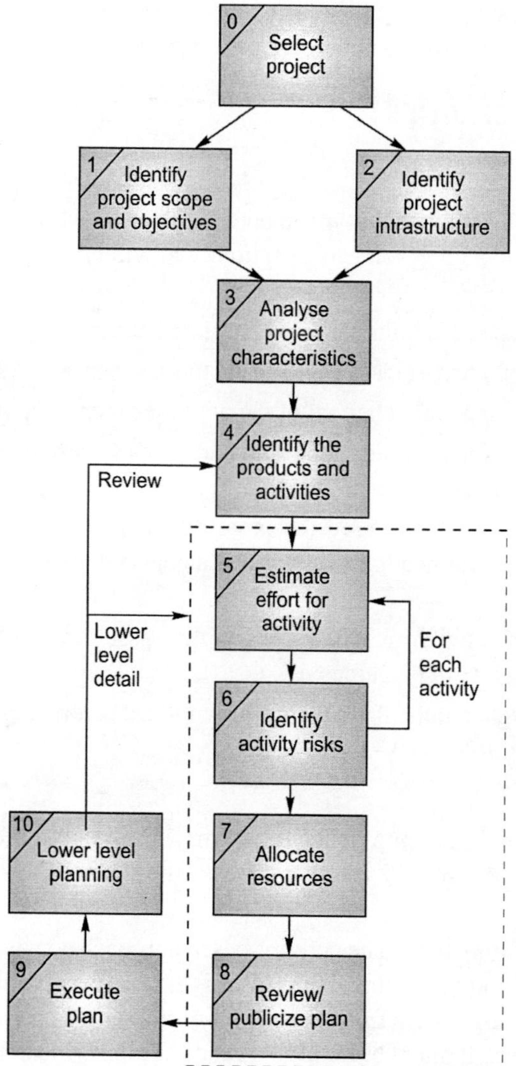

Fig. 4.1 Activity planning is carried out in Steps 4 and 5

The first step in producing the plan is to decide what activities need to be carried out and in what order they are to be done. From this we can construct an *ideal activity plan* – that is, a plan of when each activity would ideally be undertaken were resources not a constraint. It is the creation of the ideal activity plan that we shall discuss in this chapter. This activity plan is generated by Steps 4 and 5 of Step Wise (Fig. 4.1).

The ideal activity plan will then be the subject of an activity risk analysis, aimed at identifying potential problems. This might suggest alterations to the ideal activity plan and will almost certainly have implications for resource allocation.

The third step is resource allocation. The expected availability of resources might place constraints on when certain activities can be carried out, and our ideal plan might need to be adapted to take account of this.

The final step is *schedule production.* Once resources have been allocated to each activity, we will be in a position to draw up and publish a project schedule, which indicates planned start and completion dates and a resource requirements statement for each activity.

4.5 PROJECTS AND ACTIVITIES *(MDU 2012-13, 2013-14)* *(08 Marks)*

Defining activities

Before we try to identify the activities that make up a project it is worth reviewing what we mean by a project and its activities and adding some assumptions that will be relevant when we start to produce an activity plan.

- a project is composed of a number of inter-related activities;
- a project may start when at least one of its activities is ready to start;
- a project will be completed when all of the activities it encompasses have been completed;
- an activity must have a clearly defined start and a clearly defined end-point, normally marked by the production of a tangible deliverable;
- if an activity requires a resource (as most do) then that resource requirement must be forecastable and is assumed to be required at a constant level throughout the duration of the activity;
- the duration of an activity must be forecastable – assuming normal circumstances, and the reasonable availability of resources;
- some activities might require that others are completed before they can begin (these are known as *precedence requirements*).

Identifying activities

Essentially there are three approaches to identifying the activities or tasks that make up a project – we shall call them the *activity-based approach,* the *product-based approach* and the *hybrid approach.*

The activity-based approach. The activity-based approach consists of creating a list of all the activities that the project is thought to involve. This might involve a brainstorming session involving the whole project team or it might stem from an analysts of similar past projects. When listing activities, particularly for a large project, it might be helpful to subdivide the project into the main life style stages and consider each of these separately.

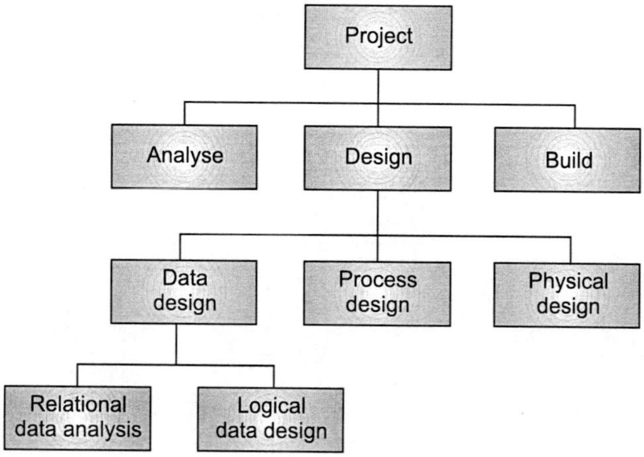

Fig. 4.2 A fragment of an activity-based Work Breakdown Structure.

Rather than doing this in an ad hoc manner, with the obvious risks of omitting or double-counting tasks, a much favoured way of generating a task list is to create a *Work Breakdown Structure* (WBS). This involves identifying the main (or high-level) tasks required to complete a project and then breaking each of these down into a set of lower-level tasks. Fig. 4.2 shows a fragment of a WBS where the design task has been broken down into three tasks and one of these has been further decomposed into two tasks.

Activities are added to a branch in the structure if they directly contribute to the task immediately above – if they do not contribute to the parent task, then they should not be added to that branch. The tasks at each level in any branch should include everything that is required to complete the task at the higher level – if they are not a comprehensive definition of the parent task, then something is missing.

When preparing a WBS, consideration must be given to the final level of detail or depth of the structure. Too great a depth will result in a large number of small tasks that will be difficult to manage, whereas a too shallow structure will provide insufficient detail for project control. Each branch should, however, be broken down at least to a level where each leaf may be assigned to an individual or responsible section within the organization.

Advantages claimed for the WBS approach include the belief that it is much more likely to result in a task catalogue that is complete and is composed of non-overlapping activities. Note that it is only the leaves of the structure that comprise the list of activities comprising the project – higher-level nodes merely represent collections of activities.

The WBS also represents a structure that may be refined as the project proceeds. In the early part of a project we might use a relatively high-level or shallow WBS, which can be developed as information becomes available, typically during the project's analysis and specification phases.

Once the project's activities have been identified (whether or not by using a WBS) they need to be sequenced in the sense of deciding which activities need to be completed before others can start.

The product-based approach. It consists of producing a Product Breakdown Structure and a Product Flow Diagram. The PFD indicates, for each product, which other products are required as inputs. The PFD can therefore be easily transformed into an ordered list of activities by identifying the transformations that turn some products into others. Proponents of this approach claim that it is less likely that a product will be left out of a PBS than that an activity might be omitted from an unstructured activity list.

This approach is particularly appropriate if using a methodology such as SSADM, which clearly specifies, for each step or task, each of the products required and the activities required to produce it. The SSADM Reference Manual provides a set of generic PBSs for each stage in SSADM (such as that shown in Fig. 4.3), which can be used as a basis for generating a project-specific PBS.

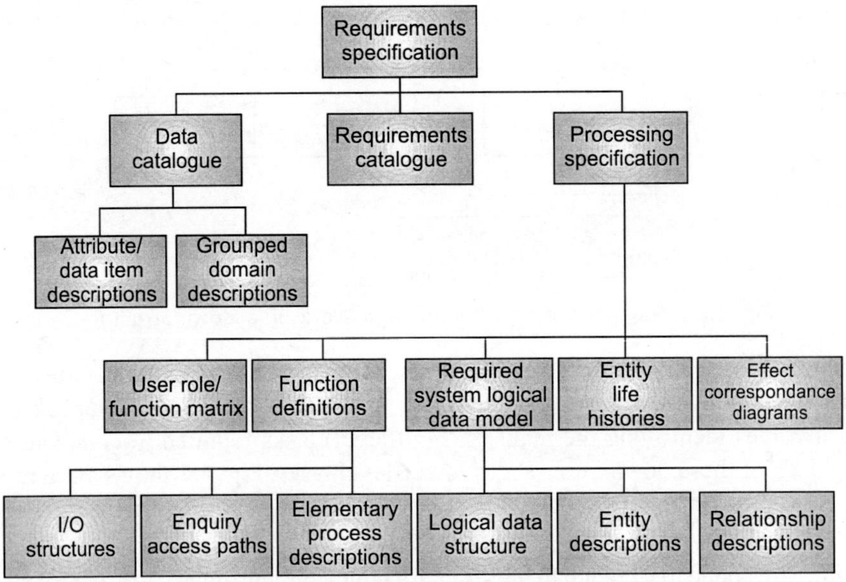

Fig. 4.3 SSADM Product Breakdown Structure for Requirements Specification (adapted from Goodland and Slater),

The SSADM Reference Manual also supplies generic activity networks and, using the project-specific PBS and derived PFD, these may be used as a basis for developing a project-specific activity network. Fig. 4.4 illustrates an activity network for the activities required to create the products in Fig. 4.3.

Notice how the development of a PFD leads directly to an activity network that indicates the sequencing of activities – in Fig. 4.4, activity 340 (Enhance required data model) requires products from activity 330 and activity 360 needs products from both activities 330 and 340.

The hybrid approach. The WBS illustrated in Fig. 4.2 is based entirely on a structuring of activities. Alternatively, and perhaps more commonly, a WBS may be based upon the project's products as illustrated in Fig. 4.5, which is in turn based on a simple list of final deliverables and, for each deliverable, a set of activities required to produce that product. Fig. 4.5 illustrates a flat WBS and it is likely that, in a project of any size, it would be beneficial to introduce additional levels – structuring both products and activities. The degree to which the structuring is product-based or activity-based might be influenced by the nature of the project and the particular development method adopted. As with a purely activity-based WBS, having identified the activities we are then left with the task of sequencing them.

A framework dictating the number of levels and the nature of each level in the structure may be imposed on a WBS. For example, in their MITP methodology, IBM recommend that the following five levels should be used in a WBS:

- **Level 1: Project.**
- **Level 2: Deliverables** such as software, manuals and training courses.
- **Level 3: Components** which are the key work items needed to produce deliverables, such as the modules and tests required to produce the system software.

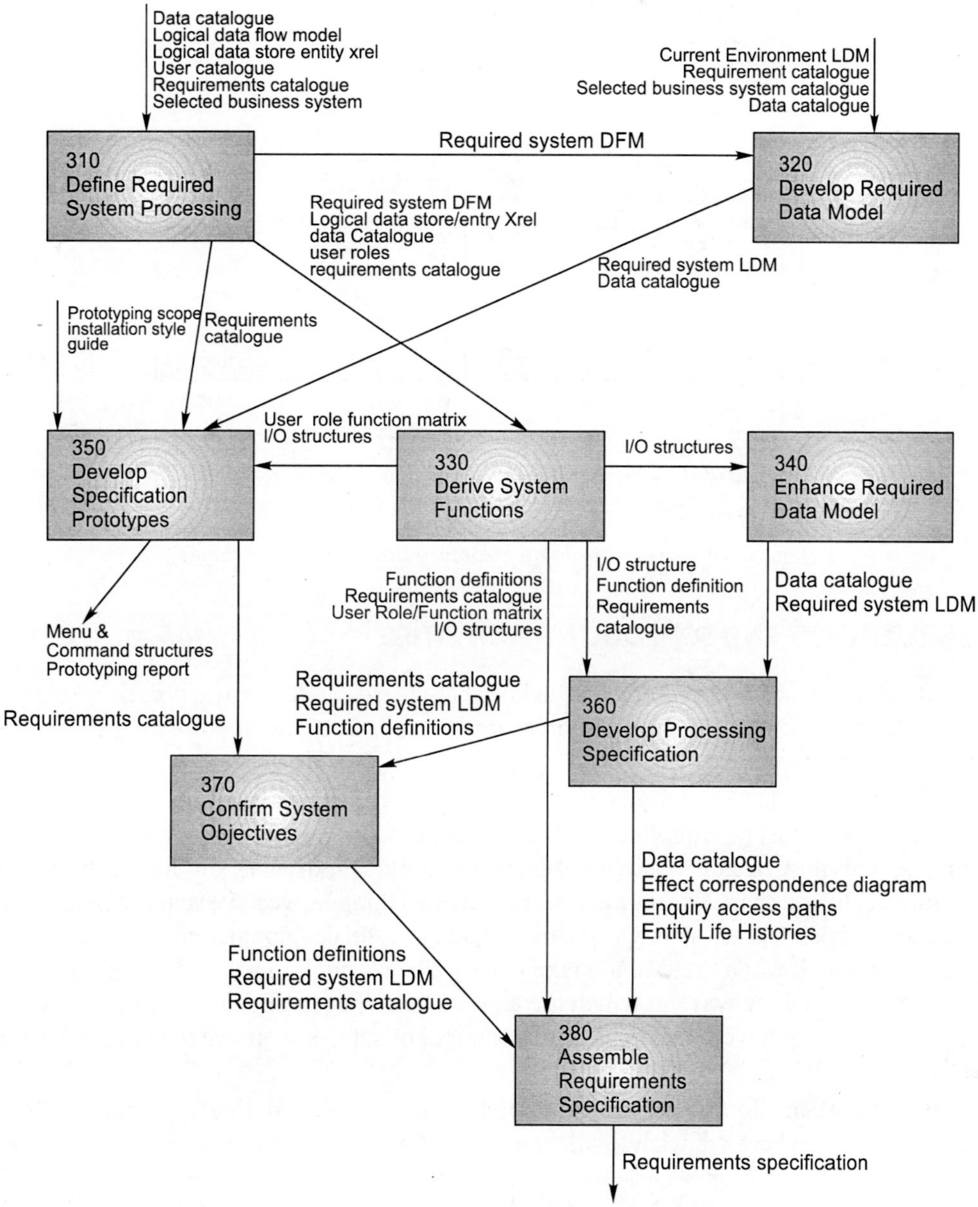

Fig. 4.4 A structuring of activities for the SSADM Requirements Specification stage [from Goodland and Slater].

- **Level 4: Work-packages** which are major work items, or collections of related tasks, required to produce a component.
- **Level 5:** Tasks which are tasks that will normally be the responsibility of a single person.

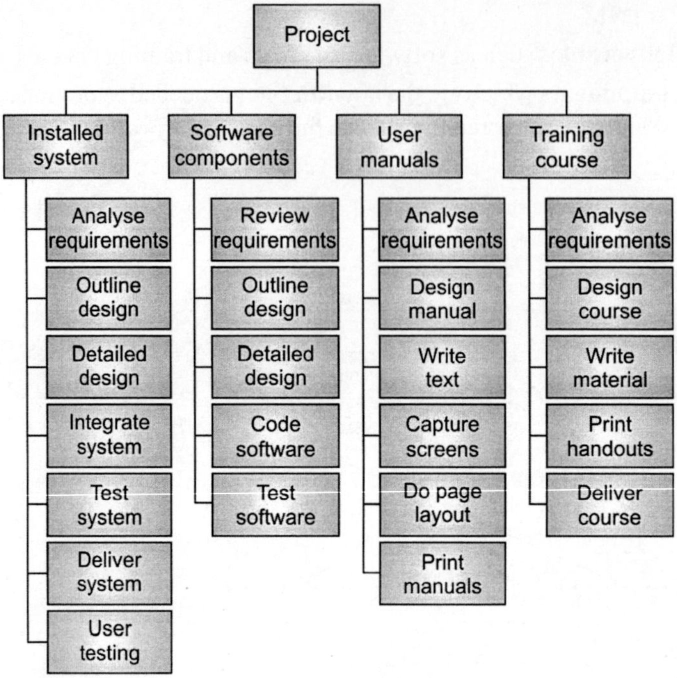

Fig. 4.5 A work breakdown structure based on deliverables.

4.6 SEQUENCING AND SCHEDULING ACTIVITIES *(MDU 2012-13) (10 Marks)*

Throughout a project, we will require a schedule that clearly indicates when each of the project's activities is planned to occur and what resources it will need. One way of presenting such a plan is to use a bar chart as shown in Fig. 4.6.

The chart shown has been drawn up taking account of the nature of the development process (that is, certain tasks must be completed before others may start) and the resources that are available (for example, activity C follows activity B because Andy cannot work on both tasks at the same time). In drawing up the chart, we have therefore done two things – we have sequenced the tasks (that is, identified the dependencies among activities dictated by the development process) and scheduled them (that is, specified when they should take place). The scheduling has had to take account of the availability of staff and the ways in which the activities have been allocated to them. The schedule might look quite different were there a different number of staff or were we to allocate the activities differently.

In the case of small projects, this combined sequencing–scheduling approach might be quite suitable, particularly where we wish to allocate individuals to particular tasks at an early planning stage. However, on larger projects it is better to separate out these two activities: to sequence the task according to their logical relationships and then to schedule them taking into account resources and other factors.

Approaches to scheduling that achieve this separation between the *logical* and the *physical* use networks to model the project and it is these approaches that we will consider in subsequent sections of this chapter.

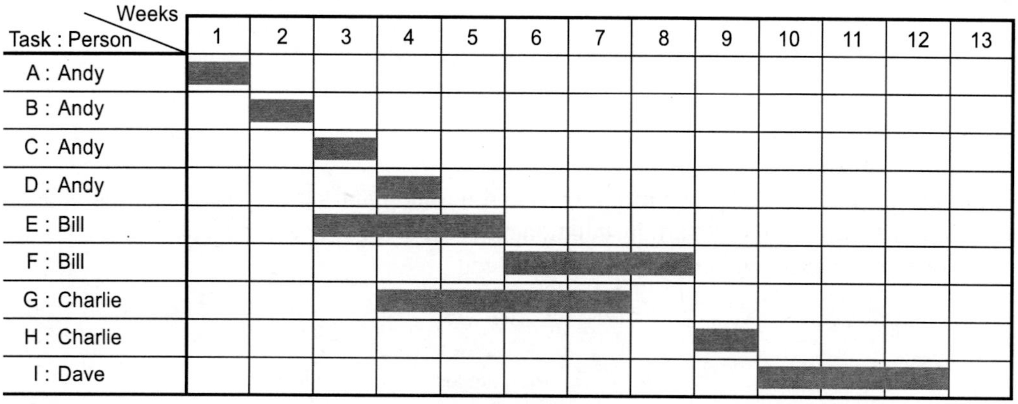

Activity key:

A : Overall design
B : Specify module 1
C : Specify module 2
D : Specify module 3
E : Code module 1
F : Code module 3
G : Code module 2
H : Integration testing
I : System testing

Fig. 4.6 A project plan as a bar chart.

4.7 NETWORK PLANNING MODELS

These project scheduling techniques model the project's activities and their relationships as a network. In the network, time flows from left to right. These techniques were originally developed in the 1950s – the two best known being CPM (Critical Path Method) and PERT (Program Evaluation Review Technique). More recently a variation on these techniques, called *precedence networks*, has become popular and it is this method that is adopted in the majority of computer applications currently available. All three methods are very similar and it must be admitted that many people use the same name (particularly CPM) indiscriminately to refer to any or all of the methods.

4.8 FORMULATING A NETWORK MODEL

(MDU 2010-11, 2012-13) (10 Marks)

The first stage in creating a network model is to represent the activities and their interrelationships as a graph. In CPM we do this by representing activities as links (arrowed lines) in the graph – the nodes (circles) representing the events of activities starting and finishing.

Constructing CPM networks

Before we look at how CPM networks are used, it is worth spending a few moments considering the rules for their construction.

A project network may have only one start node. All activities coming from that node may start immediately resources are available – that is, they do not have to wait for any other activities to be completed.

A project network may have only one end node. The end node designates the completion of the project and a project may only finish once!

A link has duration. A link represents an activity and, in general, activities take time to execute. The links are not drawn in any way to represent the activity durations. The network drawing merely represents the logic of the project – the rules governing the order in which activities are to be carried out.

Nodes have no duration. Nodes are events and, as such, are instantaneous points in time. The *Source node* is the event of the project becoming ready to start and the *sink node* is the event of the project becoming completed. Intermediate nodes represent two simultaneous events – the event of all activities leading in to a node having been completed and the event of all activities leading out of that node being in a position to be started.

In Fig. 4.7 node 3 is the event that both coding and data take-on have been completed and activity program testing is free to start. Installation may be started only when event 4 has been achieved, that is, as soon as program testing has been completed.

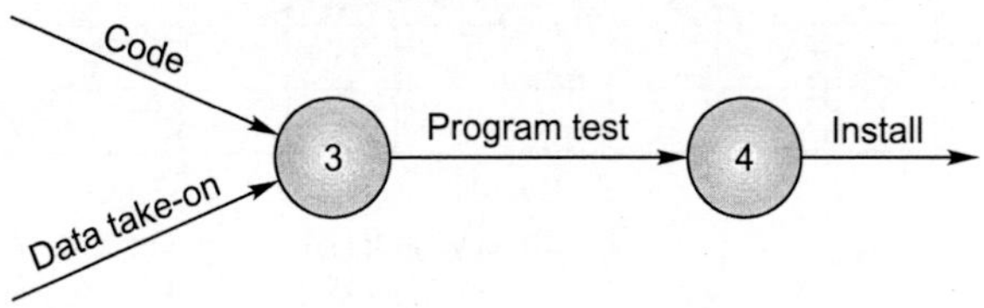

Fig. 4.7 Fragment of a CPM network.

Time moves from left to right. If at all possible, networks are drawn so that time moves from left to right. It is rare that this convention needs to be flouted but, in any case, the arrows on the activity lines give a strong visual indication of the time flow of the project.

Nodes are numbered sequentially. There are no precise rules about node numbering but nodes should be numbered so that head nodes (those at the 'arrow' end of an activity) always have a higher number than tail events (those at the 'non-arrow' end of an activity. This convention makes it easy to spot loops.

A network may not contain loops. Fig. 4.8 demonstrates a loop in a CPM network. A loop is an in error in that it represents a situation that cannot occur in practice. While loops, in the sense of iteration, may occur in practice, they cannot be directly represented in a project network. Note that the logic of Figure 4.8 suggests that program testing cannot start until the errors have been corrected.

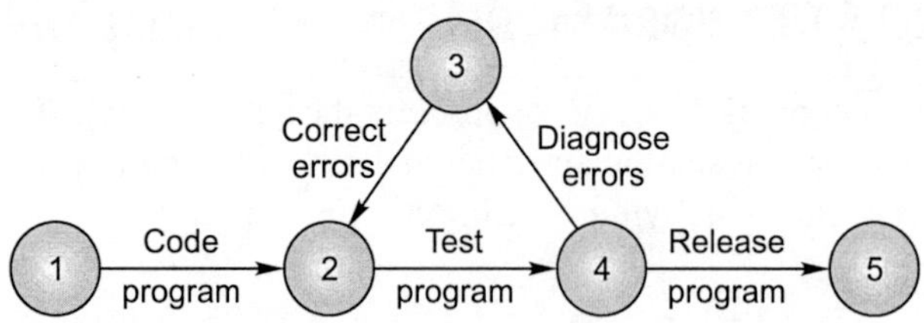

Fig. 4.8 A loop represents an impossible sequence.

If we know the number of times we expect to repeat a set of activities, a test-diagnose-correct sequence, for example, then we can draw that set of activities as a straight sequence, repeating it the appropriate number of times. If we do not know how many times a sequence is going to be repeated then we cannot calculate the duration of the project unless we adopt an alternative strategy such as redefining the complete sequence as a single activity and estimating how long it will take to complete it.

A network may not contain dangles. A dangling activity such as *Write user manual* in Fig. 4.9 cannot exist, as it would suggest there are two completion points for the project. If, in Fig. 4.9 node 5 represents the true project completion point and there are no activities dependent on activity

Write user manual, then the network should be redrawn so that activity *Write user manual* starts at node 2 and terminates at node 5 – in practice, we would need to insert a dummy activity between nodes 3 and 5. In other words, all events, except the first and the last, must have at least one activity entering them and at least one activity leaving them and all activities must start and end with an event.

Precedents are the immediate preceding activities In Fig. 4.7, the activity *Program test* cannot start until both *Code* and *Data take-on* have been completed and activity *Install* cannot start until *Program test* has finished. *Code* and *Data take-on* can therefore be said to be precedents of *Program test,* and *Program test* is a precedent of *Install.* Note that we do not speak of *Code* and *Data take-on* as precedents of *Install* – that relationship is implicit in the previous statement.

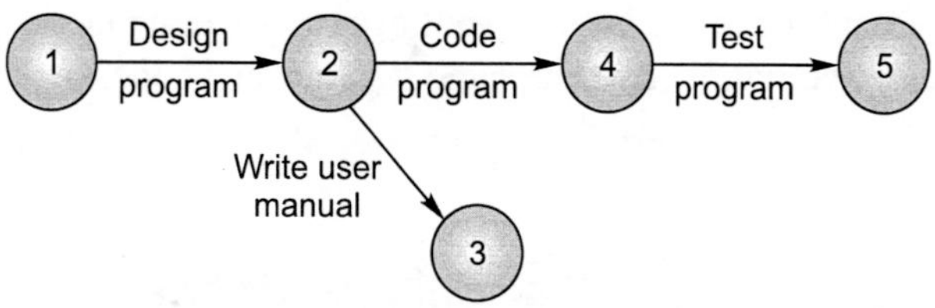

Fig. 4.9 A dangle.

4.9 USING DUMMY ACTIVITIES

When two paths within a network have a common event although they are, in other respects independent, a logical error such as that illustrated in Fig. 4.10 might occur.

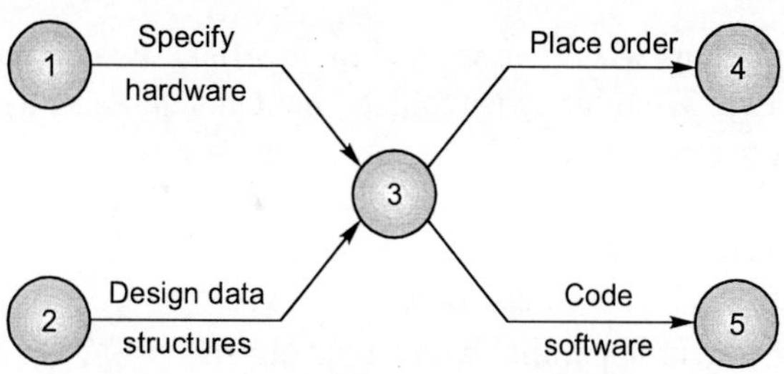

Fig. 4.10 Two paths with a common node.

Suppose that, in a particular project, it is necessary to specify a certain piece of hardware before placing an order for it and before coding the software. Before coding the software it is also necessary to specify the appropriate data structures, although clearly we do not need to wait for this to be done before the hardware is ordered.

Fig. 4.10 is an attempt to model the situation described above, although it is incorrect in that it requires both hardware specification and data structure design to be completed before either an order may be placed or software coding may commence.

We can resolve this problem by separating the two (more or less) independent paths and introducing a dummy activity to link the completion of data structure design to the start of the activity placing an order. This effectively breaks the link between data structure design and placing the order and is shown in Fig. 4.11.

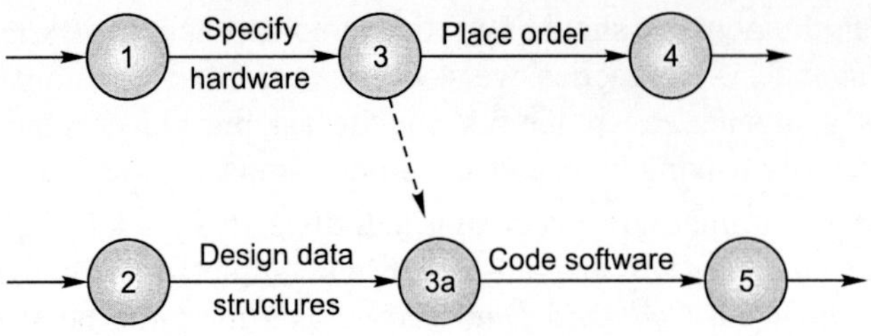

Fig. 4.11 Two paths linked by a dummy activity.

Dummy activities, shown as dotted lines and the network diagram, have a zero duration and use no resources. They are often used to aid in the layout of network drawings as in Fig. 4.12. The use of a dummy activity where two activities share the same start and end nodes makes it easier to distinguish the activity end-points.

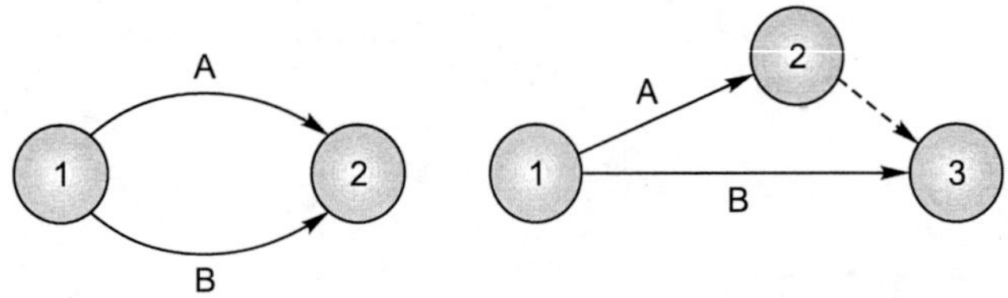

Fig. 4.12 Another use of a dummy activity.

4.10 REPRESENTING LAGGED ACTIVITIES

We might come across situations where we wished to undertake two activities in parallel so long as there is a lag between the two. We might wish to document amendments to a program as it was being tested – particularly if evaluating a prototype. In such a case we could designate an activity 'test and document amendments'. This would, however, make it impossible to show amendment recording starting after testing had begun and finishing a little after the completion of testing.

Where activities can occur in parallel with a time lag between them we represent these with pairs of dummy activities as shown in Fig. 4.13. Where the activities are lagged because a stage in one activity must be completed before the other may proceed, it is likely to be better to show each stage as a separate activity.

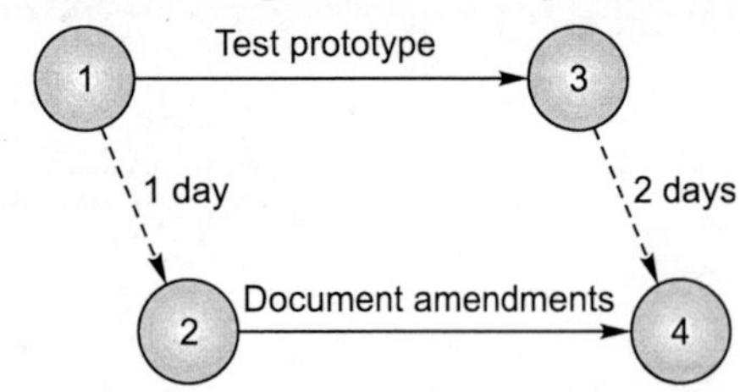

Fig. 4.13 Using the ladder technique to indicate lags.

4.11 ADDING THE TIME DIMENSION

Having created the logical network model indicating what needs to be done and the interrelationships between those activities, we are now ready to start thinking about when each activity should be undertaken.

The critical path method is concerned with two primary objectives: planning the project in such a way that it is completed as quickly as possible; and identifying those activities where a delay in their execution is likely to affect the overall end date of the project or later activities' start dates.

The method requires that for each activity we have an estimate of its duration. The network is then analysed by carrying out a *forward pass*, to calculate the earliest dates at which activities may commence and the project be completed, and a *backward pass*, to calculate the latest start dates for activities and the *critical path.*

In practice we would use a software application to carry out these calculations for anything hut the smallest of projects. It is important, though, that we understand how the calculations are carried out in order to interpret the results correctly and understand the limitations of the method.

The description and example that follow use the same example project outlined in Table 4.1 – a project composed of eight activities whose durations have been estimated as shown in the table.

Table 4.1: An example project specification with estimated activity durations and procedence requirements

Activity	*Duration (weeks)*	*Precedents*
A. Hardware selection	6	
B. Software design	4	
C. Install hardware	3	A
D. Code and test software	4	B
E. File take-on	3	B
F. Write user manuals	10	
G. User training	3	E, F
H. Install and test system	2	C, D

CPM conventions

There are a number of differing conventions that have been adopted for entering information on a CPM network. Typically the diagram is used to record information about the events rather than the activities – activity-based information (other than labels or descriptions) is generally held on a separate activity table.

One of the more common conventions for labelling nodes, and the one adopted here, is to divide the node circle into quadrants and use those quadrants to show the event number, the latest and earliest dates by which the event should occur, and the event slack (which will be explained later).

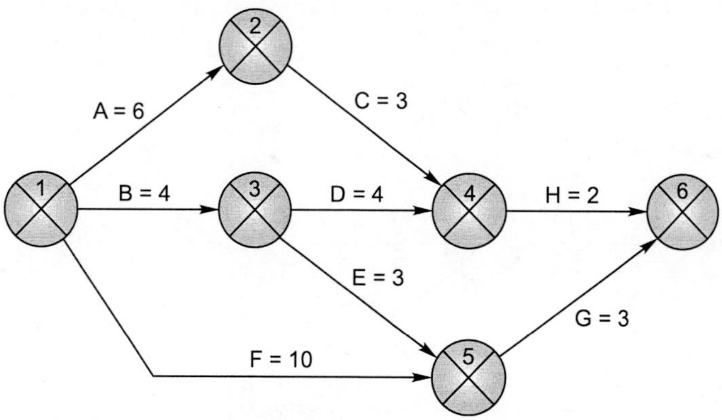

Fig. 4.14. The CPM network for the example project.

Figure 4.14 illustrates the network for the project specified in Table 4.1.

4.12 THE FORWARD PASS *(MDU 2012-13) (05 Marks)*

The forward pass is carried out to calculate the earliest data on which each event may be achieved and the earliest date on which each activity may be started and completed. The earliest date for an event is the earliest date by which all activities upon which it depends can be completed.

By convention, dates indicated the end of the period and the project is therefore shown as starting in week zero (or the beginning of week 1).

The forward pass and the calculation of earliest start dates is calculated according to the following reasoning:

- Activities A, B and F may start immediately, so the earliest date for event I is zero and the earliest start date for these three activities is also zero.
- Activity A will take 6 weeks, so the earliest it can finish is week 6 (recorded in the activity table). Therefore the earliest we can achieve event 2 is week 6.
- Activity B will take 4 weeks, so the earliest it can finish and the earliest we can achieve event 3 is week 4.
- Activity F will take 10 weeks, so the earliest it can finish is week 10 – we cannot, however, tell whether or not this is also the earliest date that we can achieve event 5 since we have not, as yet, calculated when activity E will finish.
- Activity E can start as early as week 4 (the earliest date for event 3) and, since it is forecasted to take 3 weeks, will be completed, at the earliest, at the end of week 7.
- Event 5 may be achieved when both E and F have been completed, that is, week 10 (the later of 7 and 10).
- Similarly we can reason that event 4 will have an earliest date of week 9. This is the later of the earliest finish for activity D (week 8) and the earliest finish for activity C (week 9).
- The earliest date for the completion of the project, event 6, is therefore the end of week 13 – the later of 11 (the earliest finish for H) and 13 (the earliest finish for G).

The results of the forward pass are shown in Fig. 4.15 and Table 4.2.

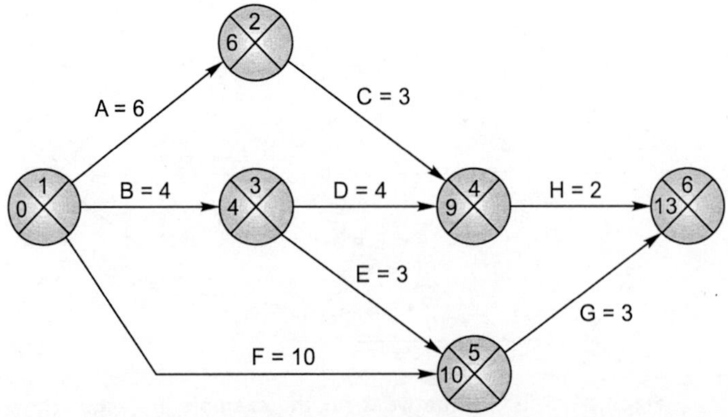

Fig. 4.15 A CPM network after the forward pass.

Table 4.2: The activity table after the forward pass

Activity	*Duration (weeks)*	*Earliest start date*	*Latest start date*	*Earliest finish date*	*Latest finish date*	*Total float*
A	6	0		6		
B	4	0		4		
C	3	6		9		
D	4	4		8		
E	3	4		7		
F	10	0		10		
G	3	10		13		
H	2	9		11		

4.13 THE BACKWARD PASS

(MDU 2012-13) (05 Marks)

The second stage is to carry out a backward pass to calculate the latest dale at which each event may be achieved, and each activity started and finished, without delaying the end date of the project. The latest date for an event is the latest date by which all immediately following activities must be started for the project to be completed on time. In calculating the latest dates, we assume that the latest finish date for the project is the same as the earliest finish date – that is, we wish to complete the project as early as possible.

Fig. 4.16 illustrates our network and Table 4.3 the activity table after carrying out the backward pass—as with the forward pass, event dates are recorded on the diagram and activity dates on the activity table.

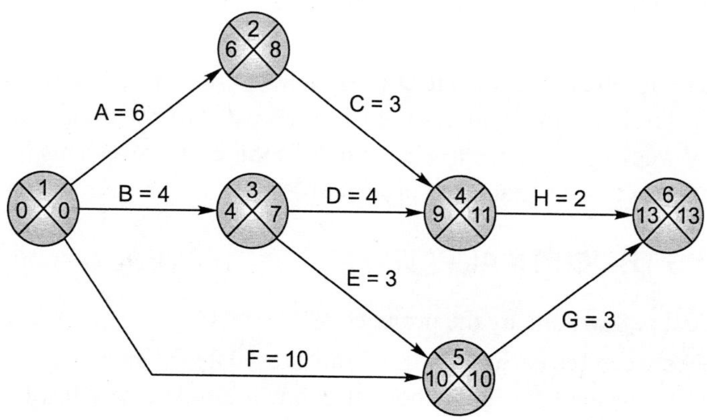

Fig. 4.16 The CPM network after the backward pass.

Table 4.3 The activity table following the backward pass

Activity	*Duration (weeks)*	*Earliest start date*	*Latest start date*	*Earliest finish date*	*Latest finish date*	*Total float*
A	6	0	2	6	8	
B	4	0	3	4	7	
C	3	6	8	9	11	
D	4	4	7	8	11	
E	3	4	7	7	10	
F	10	0	0	10	10	
G	3	10	10	13	13	
H	2	9	11	11	13	

The latest event dates are calculated as follows:

- The latest date for node 6 is assumed to be week 13, the same as the earliest date.
- The latest date for event 5 is week 10, since activity G will take 3 weeks and must be completed by week 13 if the project end date is not to be exceeded.
- The latest date for event 4 is week 11 since activity H does not need to be started until week 11 if it takes 2 weeks and does not need to be completed until week 13.
- The latest date for event 3 is the latest date by which we must be in a position to start both activities D and E. Activity E need not finish until week 10 and need not therefore start until week 7. Activity D need not finish until week 11 and, having a duration of 4 weeks, need not start until week 7. The latest date for event 3 is therefore week 7.
- The latest date for event 2 is week 8 since C, which takes 3 weeks, need not be finished until week 11.
- The earliest and latest dates for the start event must always be the same unless an arithmetic error has occurred.
- The latest date for event 1 is the latest by which we must be in a position to start activity A (which must start by week 2), activity B (which must start by week 3) and activity F (which must start by week 0). This event's latest date is therefore zero. This is, of course, not very surprising since it tells us that if the project does not start on time it won't finish on time.

4.14 IDENTIFYING THE CRITICAL PATH

(MDU 2008-09, 2011-12) (10 Marks)

Any delay on the critical path will delay the project.

The difference between the earliest date and the latest date for an event is known as the *slack* – it is a measure of how late an event may be without affecting the end date of the project. Any event with a slack of zero is critical in the sense that any delay in achieving that event will delay the completion date of the project as a whole. There will always be at least one path through the network joining those critical events - this path is known as the *critical path* (Fig. 4.17).

> **Important Note:** *The critical path is the longest path through the network.*

The significance of the critical path is two-fold.

- In managing the project, we must pay particular attention to monitoring activities on the critical path so that the effects of any delay or resource unavailability are detected and corrected at the earliest opportunity.

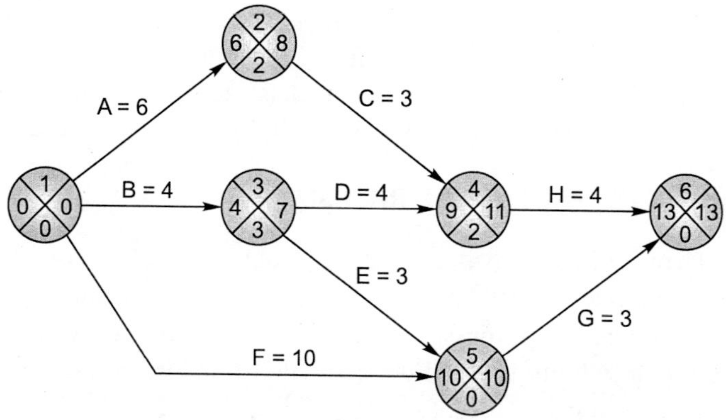

Fig. 4.17 The critical path.

- In planning the project, it is the critical path that we must shorten if we are to reduce the overall duration of the project.

4.15 ACTIVITY FLOAT

Whereas events have slack, activities posses *float*. The total float shown in Table 4.4 is the difference between the earliest start date of an activity and its latest start (or the difference between its earliest finish and its latest finish). It tells us by how long the activity's start or completion may be delayed without affecting the end date of the project.

Table 4.4. The activity schedule that wing total float for each activity

Activity	*Duration (weeks)*	*Earliest start date*	*Latest start date*	*Earliest finish date*	*Latest finish date*	*Total float*
A	6	0	2	6	8	2
B	4	0	3	4	7	3
C	3	6	8	9	11	2
D	4	4	7	8	11	3
E	3	4	7	7	10	3
F	10	0	0	10	10	0
G	3	10	10	13	13	0
H	2	9	11	11	13	2

Although the total float is shown for each activity, it really 'belongs' to a path through the network. Activities A and C each have 2 weeks total float. If, however, activity A uses up its float (that is, it is not completed until week 8) then activity B will have zero float (it will have become critical). In such circumstances it may be misleading and detrimental to the project's success to publicize total float!

There are a number of other measures of activity float, including the following:

- **Free float.** The time by which an activity may be delated without affecting any subsequent activity. It is calculated as the difference between the earliest completion date for the activity and the earliest start date of the succeeding activity. This might be considered a more satisfactory measure of float for publicizing to the staff involved in undertaking the activities.
- **Interfering float.** The difference between total float and free float. This is quite commonly used, particularly in association with the free float. Once the free float has been used (or if it is zero), the interfering float tells us by how much the activity may be delayed without delaying the project end date—even though it will delay the start of subsequent activities.

4.16 SHORTENING THE PROJECT DURATION

If we wish to shorten the overall duration of a project we would normally consider attempting to reduce activity durations. In many cases this can be done by applying more resources to the task – working overtime or procuring additional staff, for example. The critical path indicates where we must look to save time – if we are trying to bring forward the end date of the project, there is clearly no point in attempting to shorten non-critical activities. Referring to Fig. 4.14 can be seen that we could complete the project in week 12 by reducing the duration of activity F by one week (to 9 weeks).

As we reduce activity times along the critical path we must continually check for any new critical path emerging and redirect our attention where necessary.

There will come a point when we can no longer safely, or cost-effectively; reduce critical activity durations in an attempt to bring forward the project end date. Further savings, if needed, must be sought in a consideration of our work methods and by questioning the logical sequencing of activities. Generally, time savings are to be found by increasing the amount of parallelism in the network and the removal of bottlenecks (subject always, of course, to resource and quality constraints).

4.17 IDENTIFYING CRITICAL ACTIVITIES

The critical path identifies those activities which are critical to the end date of the project; however, activities that are not on the critical path may become critical. As the project proceeds, activities will invariably use up some of their float and this will require a periodic recalculation of the network. As soon as the activities along a particular path use up their total float then that path will become a critical path and a number of hitherto non-critical activities will suddenly become critical.

It is therefore common practice to identify 'near-critical' paths – those whose lengths are within, say, 10-20% of the duration of the critical path or those with a total float of less than, say, 10% of the project's uncompleted duration.

The importance of identifying critical and near-critical activities is that it is they that are most likely to be the cause of delays in completing the project. We shall see, in the next three chapters, that identifying these activities is an important step in risk analysis, resource allocation and project monitoring.

4.18 PRECEDENCE NETWORKS *(V. Important)*

Where CPM networks use links to represent activities and nodes to represent events, precedence networks use boxes (nodes) to represent activities (sometimes known as work items) and links to represent dependencies. The boxes may carry task descriptions and duration estimates and the links may contain a duration denoting a lag between the completion of one task and the start of the next.

In the notation we use here, the items indicated in the boxes are the same as those shown on the CPM network and activity table with the addition of activity span.

Activity span is the difference between the earliest start date and the latest finish date and is a measure of the maximum time allowable for the activity.

Proponents of precedence networks claim that they are easier to draw neatly, dummy activities are virtually redundant and it is easier for people to interpret them. Figure 4.18 drawn as a precedence network. As you can see, it contains much more information than the CPM network and we do not need to keep a separate activity table. The critical path through activities F and G is shown as a heavy line.

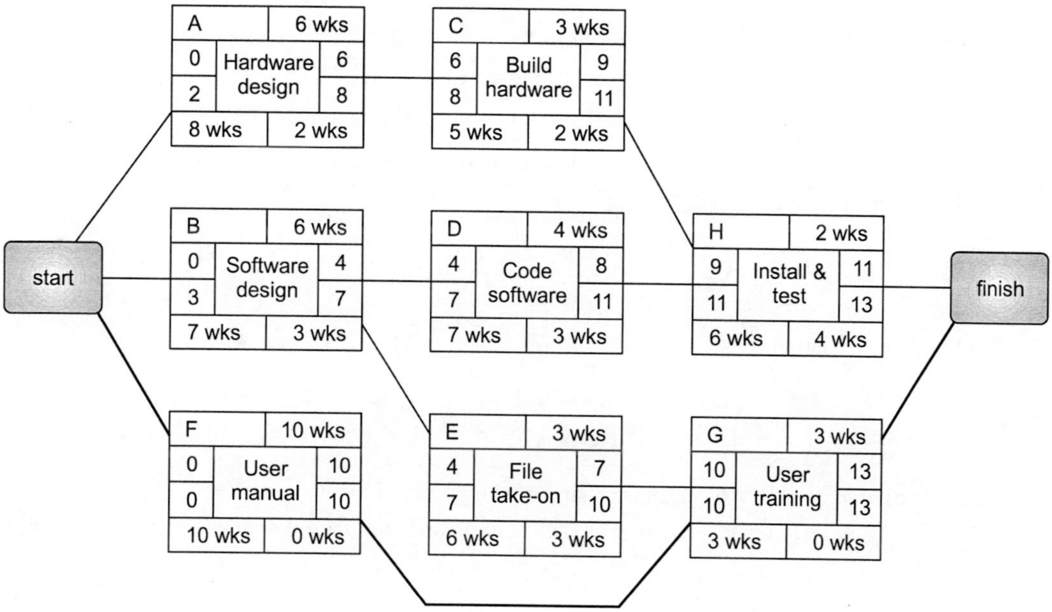

Fig. 4.18 A precedence network.

A further advantage of precedence networks is that they can represent parallel lagged activities. The example shown in Fig. 4.18, which required the use of dummy activities may be represented much more elegantly in a precedence network as shown in Fig. 4.19.

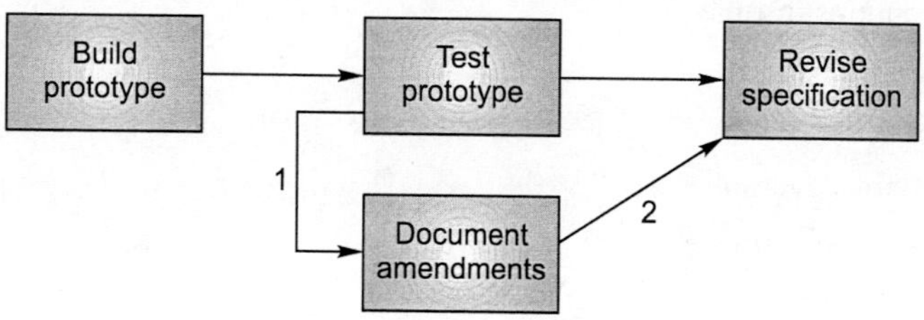

Fig. 4.19 Parallel lagged activities in a precedence network.

Analysis of precedence networks proceeds in exactly the same ways as discussed above – we carry out a forward and backward pass to calculate earliest and latest start and finish dates and identify the critical path.

CONCEPTUAL SHORT QUESTIONS WITH ANSWERS

1. What is activity float or free float?

Ans. Merge point activities that are the last activity on a stack path have what is called on activity float (or free float).

2. What is Brook's Law?

Ans. "Adding people to a late project usually only makes it later".

3. What is the element of work performed during the course of project called?

Ans. Activity.

4. Discuss about any one case tool that is used for scheduling?

Ans. A very simple CASE tool is Mac Project. It is used to generate:

(*a*) timeline chart that describes tasks as a function of chronological data.

(*b*) A Resons a Allocation Table (RAT) that contains the start and finish times of on activity.

5. Prove the following relations:

(*a*) Mean $\mu = te = \dfrac{a+4m+b}{6}$

(*b*) Standard deviation, $\sigma = \dfrac{b-a}{6}$

(*c*) Variance, $V = \left(\dfrac{b-a}{6}\right)^2$.

Where *te* σ, *a*, *m*, *b* and V have their usual meanings.

Ans. The project managers considers *a*, *m*, *b* of each activity and uses the following relations for estimations

(*a*) Mean, $\mu = te = \dfrac{a+4m+b}{6}$ ...(*i*)

(*b*) Standard deviation, $\sigma = \dfrac{b-a}{6}$ ...(*ii*)

(*c*) Variance, $V = \left(\dfrac{b-a}{2}\right)^2$ ...(*iii*)

Let '*a*' denote optimistic time.

'*b*' denote pessimistic time.

'*m*' denote mostly likely time.

'σ' denote standard deviation

't_e' denote single time estimate

'V' denote variability of an activity

Then, these can be marked in the graph given below:

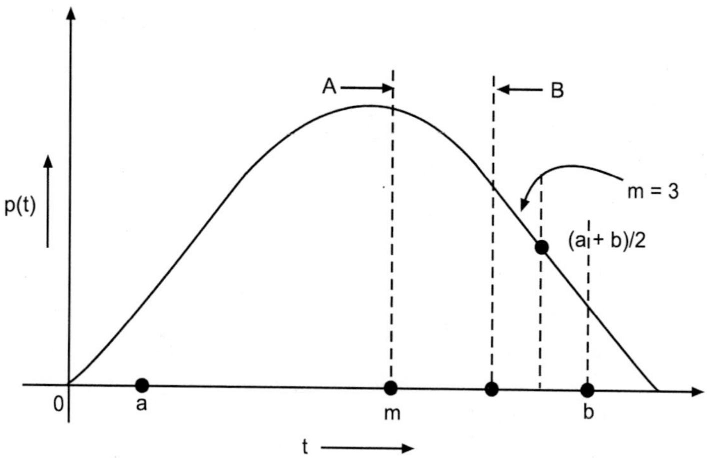

Fig. 4.20 β-curve for PERT.

Let 'A' represent mean. Also mark another point.
'B' that is midway between a and b.
∴ From the statistical characteristics of β- curve we have

$$te = m + \frac{1}{3}\text{AB} \quad \text{...}(iv)$$

But $$\text{AB} = \left(\frac{a+b}{2} - m\right) \text{ (from curve)} \quad \text{...}(v)$$

Substituting Eq. (v) in Eq. (iv), we get

$$te = m + \frac{1}{3}\left[\frac{a+b}{2} - m\right] = \frac{a+4m+b}{6} \text{ (where } a < m < b)$$

∴ $$te = \frac{a+4m+b}{6} \text{ proved.}$$

Similarly, from curve given above, we find that

$$(b-a) = \pm 3\sigma$$

or $$6\sigma = (b-a)$$

or $$\sigma = \frac{b-a}{6} \text{ proved.}$$

We also know that variance, $\text{V} = \left(\frac{b-a}{6}\right)^2$.

Now, as per the Central Limit's Theorem, the standard deviation of PERT network is the sum of standard deviations of the activities along the critical path.

∴ Variance, $$\text{V} = \sum_{i=1}^{n}\left(\frac{b-a}{6}\right)^2$$

and standard deviation of PERT network (σ) is given by the formula

$$\sigma = \sqrt{\sum_{i=1}^{n}\left(\frac{b-a}{6}\right)^2}$$

where $i \leftarrow 1$ to n and are the activities, 1, 2, 3, ..., n which are lying along the critical path.

REVIEW QUESTIONS

1. How networks charts help in scheduling a project? What is critical path?
2. Identify the critical path for the following network:

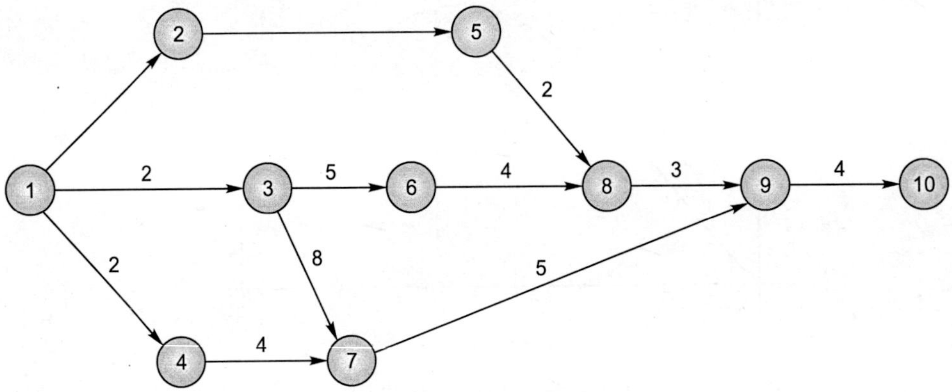

Fig. 4.21

3. Identify the critical path for the following network:

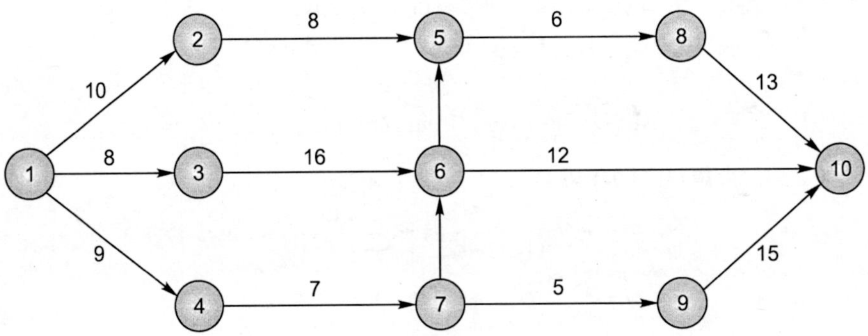

Fig. 4.22

4. Describe the network planning model and the concept of critical path.
5. Describe network chart. Explain with example, how they help in managing the schedule in a project.
6. Write short notes on following:
   (*a*) Gantt charts.
   (*b*) Product breakdown structure
   (*c*) Network planning model
   (*d*) Forward and backward pass
   (*e*) Identification of critical path
   (*f*) Precedence network.
7. Draw an activity network using either activity-on-node or activity-on-arrow network conventions for each of the following projects:
   (*a*) getting married;
   (*b*) choosing and purchasing a desktop computer;
   (*c*) organizing and carrying out a survey of user's opinions of an information system.

CHAPTER 5

RISK MANAGEMENT

INSIDE THIS CHAPTER

5.1 INTRODUCTION

Risk can be defined as the combination of the probability of an event and its consequences. **(ISO/IEC Guide 73).**

Risk management is the area that tries to ensure that the impact of the risks on cost, quality and schedule is minimal.

In this chapter, we are concerned with the risk of the development project's not proceeding according to plan. We are primarily concerned with the risks of the project's running late or over budget and with the identification of the steps that can be taken to avoid or minimize those risks.

Some risks are more important than others. Whether or not a particular risk is important depends on the nature of the risk, its likely effects on a particular activity and the criticality of the activity. High risk activities on a project's critical path are a cause for concern.

To reduce these dangers, we must ensure that risks are minimized or, at least, distributed over the project and, ideally, removed from critical path activities.

The risk of an activity's running over time is likely to depend, at least in part, on who is doing or managing it. The evaluation of risk and the allocation of staff and other resources are therefore closely connected as shown in Fig. 5.1.

Risk analysis and management are a series of steps that help a software team to understand and manage uncertainity. Many problems can plague a software project. A risk is a potential, problem – it might happen, it may not. But, regardless of the outcome, it's really good idea to identify it, assess its probability of occurence, estimate its impact, and establish a contingency plan should the actually occur.

In software engineering risk is a measure of the inability to achieve overall program objectives within defined cost, schedule, and technical constraints.

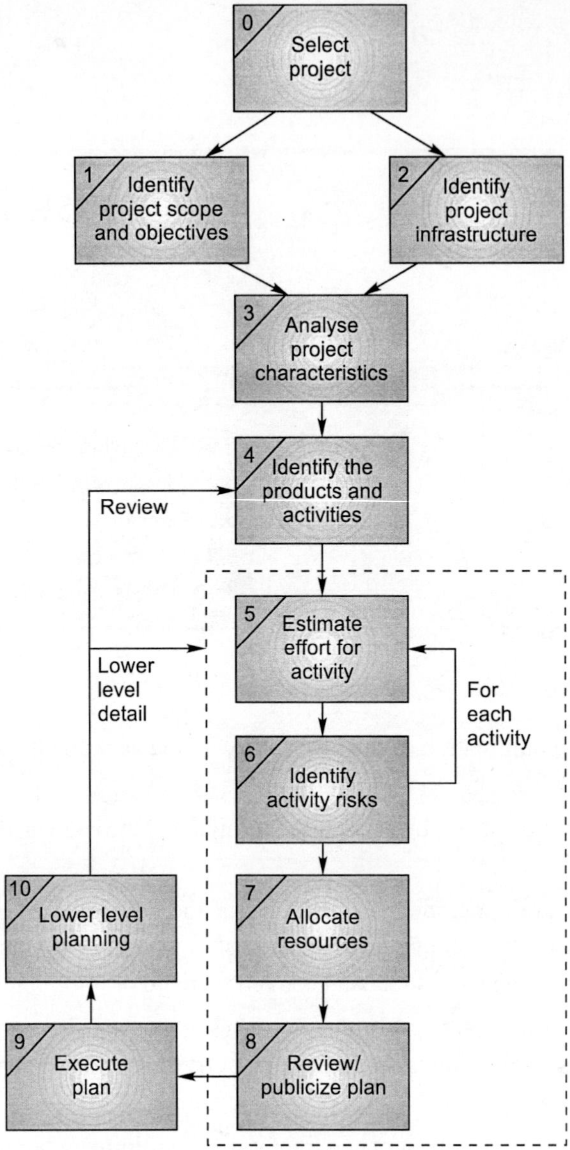

Fig. 5.1 Risk analysis is carried out in Steps 3 and 6.

5.2 THE NATURE OF RISK (*MDU 2010-11*) (*05 Marks*)

For the purpose of identifying and managing those risks that may cause a project to overrun its time-scale or budget, it is convenient to identify three types of risk:

- those caused by the inherent difficulties of estimation;
- those due to assumptions made during the planning process;
- those of unforeseen (or at least unplanned) events occurring.

Estimation errors

Some tasks are harder to estimate than others because of the lack of experience of similar tasks or because of the nature of a task. Producing a set of user manuals is reasonably straightforward and, given that we have carried out similar tasks previously, we should be able to estimate with some degree of accuracy how long it will take and how much it will cost. On the other hand, the time required for program testing and debugging, might be difficult to predict with a similar decree of accuracy – even if we have written similar programs in the past.

Estimation can be improved by analysing historic data for similar activities and similar systems. Keeping records comparing our original estimates with the final outcome will reveal the type of tasks that are difficult to estimate correctly.

Planning assumptions

At every stage during planning, assumptions are made which, if not valid, may put the plan at risk. Our activity network, for example, is likely to be built on the assumption of using a particular design methodology – which may be subsequently changed. We generally assume that, following coding, a module will be tested and then integrated with others – we might not plan for module testing showing up the need for changes in the original design but, in the event, it might happen.

At each stage in the planning process, it is important to list explicitly all of the assumptions that have been made and identify what effects they might have on the plan if they are inappropriate.

Eventualities

Some eventualities might never be foreseen and we can only resign ourselves to the fact that unimaginable things do, sometimes, happen. They are, however, very rare. The majority of unexpected events can, in fact, be identified – the requirements specification might be altered after some of the modules have been coded, the senior programmer might take maternity leave, the required hardware might not be delivered on time. Such events do happen from time to time and, although the likelihood of any one of them happening during a particular project may be relatively low, they must be considered and planned for.

5.3 MANAGING RISK *(MDU 2008-09, 2011-12, 2012-13)* *(10 Marks)*

Risk is defined as an exposure to the chance of injury or loss. That is risk implies that there is a possiblity that something negative may happen. In the context of software projects, negative implies that there is an adverse effect on cost, quality or schedule.

Risk Management is a scientific process based on the application of game theory, decision theory, probability theory and utility theory.

The objective of risk management is to avoid or minimize the adverse effects of unforeseen events by avoiding the risks or drawing up contingency plans for dealing with them.

There are a number of models for risk management, but most are similar, in that they identify two main components – *risk identification* and *risk management.* An example of an often-used model is that in Fig. 5.2, which shows a task breakdown structure for what Barry Boehm calls *risk engineering.*

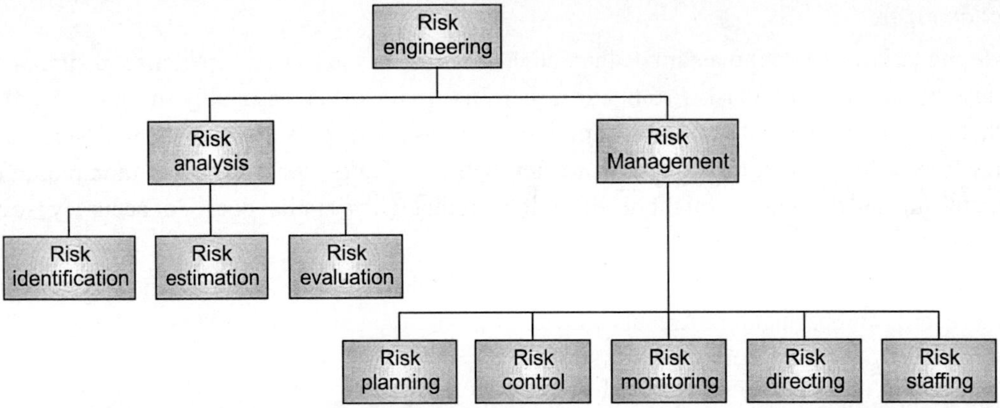

Fig. 5.2 Boehm's risk engineering task breakdown

- **Risk identification** consists of listing all of the risks that can adversely affect the successful execution of the project.
- **Risk estimation** consists of assessing the likelihood and impact of each hazard.
- **Risk evaluation** consists of ranking the risks and determining risk aversion strategies.
- **Risk planning** consists of drawing up contingency plans and, where appropriate, adding these to the project's task structure. With small projects, risk planning is likely to be the responsibility of the project manager but medium or large projects will benefit from the appointment of a full-time risk manager.
- **Risk control** concerns the main functions of the risk manager in minimising and reacting to problems throughout the project. This function will include aspects of quality control in addition to dealing with problems as they occur.
- **Risk monitoring** must be an ongoing activity, as the importance and likelihood of particular risks can change as the project proceeds.
- **Risk directing** and **risk staffing** are concerned with the day-to-day management of risk. Risk aversion and problem solving strategies frequently involve the use of additional staff and this must be planned for and directed.

Whatever task model or whichever techniques are used, risk management will not be affective unless all project staff are risk-oriented and are provided with an environment where they can freely discuss the risks that might affect a project. All too often, team members who identify potential risks at an early stage are seen as having a negative attitude.

Writing about attitudes to risk, Dwayne Phillips remarks that 'I have seen a room get suddenly quiet when someone brings up a "concern" but says that 'pretending that problems will not occur will not prevent them'. For effective risk management, it is important that the project team are encouraged to identify and discuss risks as early as possible in the project's life.

The techniques described in the rest of this chapter describe how risks can be identified and quantified and are designed to provide a framework that engenders a positive attitude to the analysis and management of project risks.

5.4 RISK IDENTIFICATION *(MDU 2010-11, 2012-13, 2013-14) (10 Marks)*

The first stage in any risk assessment exercise is to identify the hazards that might affect the duration or resource costs of the project. A hazard is an event that might occur and will, if it does occur, create a problem for the successful completion of the project. In identifying and analysing risks, we can usefully distinguish between the cause (or hazard), its immediate effect (the problem that it creates) and the risk that it will pose to the project.

For example, the illness of a team member is a *hazard* that might result in the *problem* of late delivery of a component. The late delivery of that component is likely have an effect on other activities and might, particularly if it is on the critical path, put the project completion date at *risk*.

A common way of identifying hazards is to use a checklist listing all the possible hazards and factors that influence them. Typical checklists list many, even hundreds, of factors and there are, today, a number of knowledge-based software products available to assist in this analysis.

Some hazards are *generic risks* – that is, they are relevant to all software projects and standard checklists can be used and augmented from an analysis of past projects to identify them. These will include risks such as misunderstanding the requirements or key personnel being ill. There will also be *specific risks* that are relevant to an individual project and these are likely to be more difficult to identify without an involvement of the members of the project team and a working environment that encourages risk assessment.

The categories of factors that will need to be considered include the following:

Application factors. The nature of the application – whether it is a simple data processing application, a safety-critical system or a large distributed system with real-time elements – is likely to be a critical factor. The expected size of the application is also important – the larger the system, the greater is the likelihood of errors and communication and management problems.

Staff factors. The experience and skills of the staff involved are clearly major factors – an experienced programmer is, one would hope, less likely to make errors than one with little experience. We must, however, also consider the appropriateness of the experience – experience in coding small data processing modules in Cobol may be of little value if we are developing a complex real-time control system using C++.

Such factors as the level of staff satisfaction and the staff turn-over rates are also important to the success of any project – demotivated staff or key personnel leaving unexpectedly have caused many a project to fail.

Project factors. It is important that the project and its objectives are well defined and that they are absolutely clear to all members of the project team and all key stakeholders. Any possibility that this is not the case will pose a risk to the success of the project. Similarly, an agreed and formal quality plan must be in place and adhered to by all participants and any possibility that the quality plan is inadequate or not adhered to will jeopardize the project.

Project methods. Using well specified and structured methods (such as PRINCE 2 and SSADM) for project management and system development will decrease the risk of delivering a system that is unsatisfactory or late. Using such methods for the first time, though, may cause problems and delays – it is only with experience that the benefits accrue.

Hardware/software factors. A project that requires new hardware for development is likely to pose a higher risk than one where the software can be developed on existing (and familiar) hardware. Where a system is developed on one type of hardware or software platform to be used on another there might be additional (and high) risks at installation.

Changeover factors. The need for an 'all-in-one' changeover to the new system poses particular risks. Incremental or gradual changeover minimizes the risks involved but is not always practical. Parallel running can provide a safety net but might be impossible or too costly.

Supplier factors. The extent to which a project relies on external organizations that cannot be directly controlled often influences the project's success. Delays in, for example, the installation of telephone lines or delivery of equipment may be difficult to avoid – particularly if the project is of little consequence to the external supplier.

Environment factors. Changes in the environment can affect a project's success. A significant change in the taxation regulations could, for example, have serious consequences for the development of a payroll application.

Health and safety factors. While not generally a major issue for software projects (compared, say, to civil engineering projects), the possible effects of project activities on the health and safety of the participants and the environment should be considered. BS 6079 states that 'every project should include an audit of these specific risks before work starts' and that 'audit updates should be scheduled as part of the overall project plan'.

Although some factors might influence the project as a whole, it is necessary to consider them individually for each activity – a key member of staff being ill during fact-finding might, for example, be far less serious than a similar absence during user training. Within a PRINCE 2 environment it can be appropriate to list the factors for each of the products identified in the product breakdown structure.

5.5 RISK ANALYSIS *(MDU 2007-08, 2012-13)* *(05 Marks)*

Risk Analysis is also known as Risk Quantification. Risk Quantification or risk analysis is the process of evaluating risks to assess the range of possible project outcomes. In other words we can define Risk Analysis (RA) as a process that defines activities and methods to estimate and evaluate risk.

Having identified the risks that might affect our project we need some way of assessing their importance. Some risks will be relatively unimportant (for example, the risk that some of the documentation is delivered a day late), whereas some will be of major significance (such as the risk that the software is delivered late). Some are quite likely to occur (it is quite likely, for example, that one of the software developers in a team will take a few days sick leave during a lengthy project), whereas others are relatively unlikely (hardware failure causing loss of completed code, perhaps).

The risk analysis process begins with list of risks obtained from the earlier process of risk identification. Each risk from this list is taken for estimation and evaluation. To act on this stem, the organization needs an evaluation criteria and the risk data base.

The probability of a hazard's occurring is known as the *risk likelihood;* the effect that the resulting problem will have on the project, if it occurs, is known as the *risk impact* and the importance of the risk is known as the *risk value* or *risk exposure.* The risk value is calculated as:

$$\textit{risk exposure} = \textit{risk likelihood} \times \textit{risk impact}$$

Ideally the risk impact is estimated in monetary terms and the likelihood assessed as a probability. In that case the risk exposure will represent an expected cost in the same sense that we calculated expected costs and benefits when discussing cost-benefit analysis. The risk exposures for various risks can then be compared with each other to assess the relative importance of each risk and they can be directly compared with the costs and likelihoods of success of various contingency plans.

However, estimation of these costs and probabilities is likely to be difficult, subjective, time-consuming and costly. In spite of this, it is valuable to obtain some quantitative measure of risk likelihood and impact because, without these, it is difficult to compare or rank risks in a meaningful way. Moreover, the effort put into obtaining a good quantitative estimate can provide a deeper and valuable understanding of the problem.

Many risk managers use a simple scoring method to provide a quantitative measure for assessing each risk. Some just categorize likelihoods and impacts as high, medium or low, but this form of ranking does not allow the calculation of a risk exposure. A better and popular approach is to score the likelihood and impact on a scale of, say, 1 to 10 where the hazard that is most likely to occur receives a score of 10 and the least likely a score of 1.

Ranking likelihoods and impacts on a scale of 1 to 10 is relatively easy, but most risk managers will attempt to assign scores in a more meaningful way such that, for example, a likelihood scoring 8 is considered twice as likely as one with a score of 4.

Impact measures, scored on a similar scale, must take into account the total risk to the project. This must include the following potential costs:

- the cost of delays to scheduled dates for deliverables;
- cost overruns caused by using additional or more expensive resources;
- the costs incurred or implicit in any compromise to the system's quality or functionality.

Table 5.1 illustrates part of risk value assessment. Notice that the hazard with the highest risk value might not be the one that is most likely nor the one with the greatest potential impact.

Table 5.1: Part of risk exposure assessment

	Hazard	*Likelihood*	*Impact*	*Risk exposure*
R1	Changes to requirements specification during coding	1	8	8
R2	Specification takes longer than expected	3	7	21
R3	Staff sickness affecting critical path activities	5	7	35
R4	Staff sickness affecting non-critical activities	10	3	30
R5	Module coding takes longer than expected	4	5	20
R6	Module testing demonstrates errors or deficiencies in design	1	10	10

Prioritizing the risks

Managing risk involves the use of two strategies:

- reducing the risk exposure by reducing the likelihood or impact:
- drawing up contingency plans to deal with the risk should it occur.

Any attempt to reduce a risk exposure or put a contingency plan in place will have a cost associated with it. It is therefore important to ensure that this effort is applied in the most effective way and we need a way of prioritizing the risks so that the more important ones can receive the greatest attention.

Risk exposures based on scoring methods must be treated with some caution. Risk exposure assessment is shown in Table 5.1 does not indicate, for example, that risk R5 is twice as important as R6. Nor can it be taken as necessarily meaning that R2 is more important than R5. In the first case, this is because we cannot interpret the risk exposure values quantitatively because they are based on a non-cardinal scoring method. In the second case, the exposure values are far too close for us to be able to distinguish between them – particularly in view of the somewhat approximate and subjective way in which it is likely to have assessed the likelihoods and, perhaps to a lesser extent, the impacts.

The risk exposures will, however, allow us to obtain an approximate ranking in order of importance. Considering just the risks in Table 5.1. R3 and R4 are, on this basis, clearly the most important and we could classify them as being high risk concerns. There is a significant difference between the exposure scores of these two and one with the next highest exposure, R2. R2 and R5 have similar scores and might be thought of as medium priority risks. The two remaining risks, R1 and R6 have quite low exposure values and can therefore be classified as low risk items.

In practice, there are generally other factors, in addition to the risk exposure value, that must also be taken into account when prioritizing risks.

- **Confidence of the risk assessment.** Some of our risk exposure assessments will be relatively poor. Where this is the case, there is a need for further investigation before action can be planned.
- **Compound risks.** Some risks will be dependant on others. Where this is the case, they should be treated together as a single risk.
- **The number of risks.** There is a limit to the number of risks that can be effectively considered and acted on by a project manager. We might therefore wish to limit the size of the prioritized list.
- **Cost of action.** Some risks, once recognized, can be reduced or avoided immediately with very little cost or effort and it is sensible to take action on these regardless of their risk value. For other risks we need to compare the costs of taking action with the benefits of reducing the risk. One method for doing this is to calculate the *risk reduction leverage* (RRL) using the equation

$$RRL = \frac{RE_{before} - RE_{after}}{\text{risk reduction cost}}$$

where RE_{before} is the original risk exposure value, RE_{after}, is the expected risk exposure value after taking action and the risk reduction cost is the cost of implementing the risk reduction action. Risk reduction costs must be expressed in the same units as risk values – that is, expected monetary values or score values. If the values are expected monetary values then an RRL greater than one indicates that we can expect to gain from implementing the risk reduction plan because the expected reduction in risk exposure is greater than the cost of the plan. In either case the higher the leverage value for a risk then the more worthwhile it will be to plan the risk reduction action.

5.6 REDUCING THE RISKS

(MDU 2008-09) (08 Marks)

Broadly, there are five strategies for risk reduction.

- **Hazard prevention.** Some hazards can be prevented from occurring or their likelihood reduced to insignificant levels. The risk of key staff being unavailable for meetings can be minimized by early scheduling, for example.

- **Likelihood reduction.** Some risks, while they cannot be prevented, can have their likelihoods reduced by prior planning. The risk of late changes to a requirements specification can, for example, be reduced by prototyping. Prototyping will not eliminate the risk of late changes and will need to be supplemented by contingency planning.
- **Risk avoidance.** A project can, for example, be protected from the risk of overrunning the schedule by increasing duration estimates or reducing functionality.
- **Risk transfer.** The impact of some risks can be transferred away from the project by, for example, contracting out or taking out insurance.
- **Contingency planning.** Some risks are not preventable and contingency plans will need to be drawn up to reduce the impact should the hazard occur. A project manager should draw up contingency plans for using agency programmers to minimize the impact of any unplanned absence of programming staff.

Many of these generic checklists, as well as listing common generic hazards, list typical actions for risk reduction. The checklist in Table 5.2 is based upon an often-quoted list produce by Barry Boehm.

Table 5.2: Software projects risks and strategies for risk reduction

Risk	*Risk reduction techniques*
Personnel shortfalls	staffing with top talent; job matching; team building; training and career development; early scheduling of key personnel.
Unrealistic time and cost estimates	multiple estimation techniques; design to cost; incremental development; recording and analysis of past projects; standardization of methods.
Developing the wrong software functions	improved project evaluation; formal specification methods; user surveys; prototyping; early users' manuals.
Developing the wrong user interface	prototyping; task analysis; user involvement.
Gold plating	requirements scrubbing; prototyping; cost-benefit analysis; design to cost.
Late changes to requirements	stringent change control procedures; high change threshold; incremental prototyping; incremental development (defer changes).
Shortfalls in external supplied components	benchmarking; inspections; formal specifications; contractual agreements; quality assurance procedures and certification.
Shortfalls in externally performed tasks	quality assurance procedures; competitive design or prototyping; teambuilding; contract incentives.
Real-time performance shortfalls	simulation; benchmarking; prototyping; tuning; technical analysis.
Development technically too difficult	technical analysis; cost-benefit analysis; prototyping; staff training and development.

5.7 EVALUATING RISKS TO THE SCHEDULE *(MDU 2010-11, 2012-13) (10 Marks)*

We have seen that not all risks can be eliminated – even those that are classified as avoidable or manageable can, in the event, still cause problems affecting activity durations. By identifying and categorizing those risks, and in particular, their likely effects on the duration of planned activities, we can assess what impact they are likely to have on our activity plan.

We will now take a look at two methods for assessing the effects of these uncertainties on the project schedule.

Using PERT to evaluate the effects of uncertainty

PERT was developed to take account of the uncertainty surrounding estimates of task durations. It was developed in an environment of expensive, high-risk and state-of-the-art projects - not that dissimilar to many of today's large software projects.

The method is very similar to the CPM technique (indeed many practitioners use the terms PERT and CPM interchangeably) but, instead of using a single estimate for the duration of each task, PERT requires three estimates.

- **Most likely time.** The time we would expect the task to take under normal circumstances. We shall denote this by the letter *m*.
- **Optimistic time.** The shortest time in which we could expect to complete the activity, barring outright miracles. We shall use the letter *a* to denote this.
- **Pessimistic time.** The worst possible time allowing for all reasonable eventualities but excluding 'acts of God and warfare' (as they say in most insurance exclusion clauses). We shall denote this by *b*.

PERT then combines these three estimates to form a single expected duration t_e, using the formula

$$t_e = \frac{a + 4m + b}{6}$$

Using expected durations

The expected durations are used to carry out a forward pass through a network, using the same method as the CPM technique. In this case, however, the calculated event dates are not the earliest possible dates but the dates by which we expect to achieve those events.

Table 5.3: PERT activity times estimates

Activity	*Activity durations (weeks)*		
	Optimistic (*a*)	*Most likely* (*m*)	*Pessimistic* (*b*)
A	5	6	8
B	3	4	5
C	2	3	3
D	3.5	4	5
E	1	3	4
F	8	10	15
G	2	3	4
H	2	2	2.5

The PERT network illustrated in Fig. 5.3 indicates that we expect the project to take 13.5 weeks. In Fig. 5.3 we have used an activity-on-arrow network as this form of presentation makes it easier to separate visually the estimated activity data (expected durations and, later, their standard deviations) from the calculated data (expected completion dates and target completion dates). The method can, of course, be equally well supported by activity-on-node diagrams.

Unlike the CPM approach, the PERT method does not indicates the earliest date by which we could complete the project but the expected (or most likely) date. An advantage of this approach is that it places an emphasis on the uncertainty of the real world. Rather than being tempted to say '*the completion date for the project is ...*' were are led to say '*we expected to complete the project by ...*'.

It also focuses attention on the uncertainty of the estimation of activity durations. Requesting three estimates for each activity emphasizes the fact that we are nto certain what will happen – we are forced to take into account the fact that estimates are approximate.

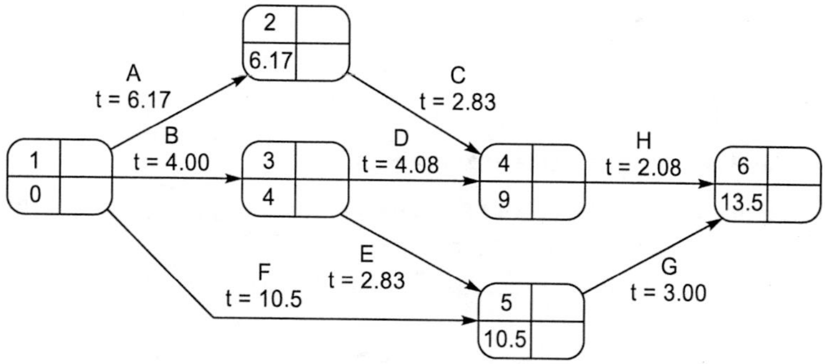

Fig. 5.3 The PERT network after the forward pass

Activity standard deviations

This standard deviation formula is based on the rationale that there are approximately six standard deviations between the extreme tails of many statistical distributions.

A quantitative measure of the degree of uncertainty of an activity duration estimate may be obtained by calculating the standard deviation s of an activity time, using the formula

$$s = \frac{b - a}{6}$$

The activity standard deviation is proportional to the difference between the optimistic and pessimistic estimates, and can be used as a ranking measure of the degree of uncertainty or risk for each activity. The activity expected durations and standard deviations for our sample project are shown in Table 5.4.

Table 5.4: Expected times and standard deviations

Activity	*Activity durations (weeks)*				
	Optimistic (a)	*Most likely* (m)	*Pessimistic* (b)	*Expected* (t_e)	*Standard deviation* (s)
A	5	6	8	6.17	0.50
B	3	4	5	4.00	0.33
C	2	3	3	2.83	0.17
D	3.5	4	5	4.08	0.25
E	1	3	4	2.83	0.50
F	8	10	15	10.50	1.17
G	2	3	4	3.00	0.33
H	2	2	2.5	2.08	0.08

The likelihood of meeting targets

The main advantage of the PERT technique is that it provides a method for estimating the probability of meeting or missing target dates. There might be only a single target date – the project completion – but we might wish to set additional intermediate targets.

Suppose that we must complete the project within 15 weeks at the outside. We expect it will take 13.5 weeks but it could take more or, perhaps, less. In addition, suppose that activity C must be completed by week 10, as it is to be carried out by a member of staff who is scheduled to be working on another project and that event 5 represents the delivery of intermediate products to the customer.

These three targets dates are shown on the PERT network in Fig. 5.4.

The PERT technique uses the following three-step method for calculating the probability of meeting or missing a target date:

- calculate the standard deviation of each project event;
- calculate the z value for each event that has a target date;
- convert z values to a probabilities.

Calculating the standard deviation of each project event ***(Important)***

Standard deviations for the project events can be calculated by carrying out a forward pass using the activity standard deviations in a manner similar to that used with expected durations. There is, however, one small difference – to add two standard deviations we must add their squares and then find the square root of the sum.

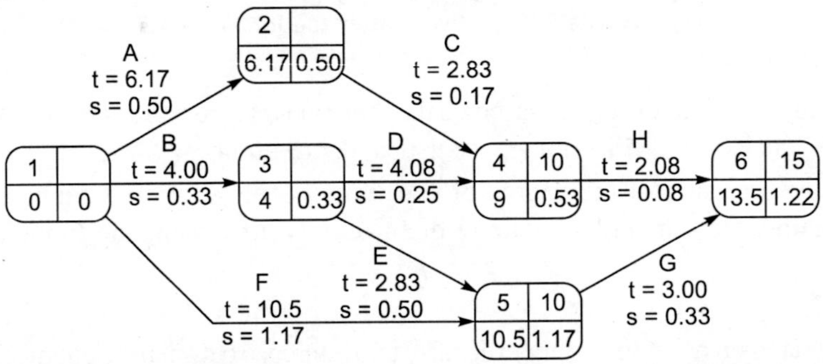

Fig. 5.4 The PERT network with three target dates and calculated event standard deviations

5.8 CALCULATING THE Z VALUES *(MDU 2013-14)* *(10 Marks)*

The z value is calculated for each node that has a target date. It is equivalent to the number of standard deviations between the node's expected and target dates. It is calculated using the formula

$$z = \frac{T - t_e}{s}$$

where t_e is the expected date and T the target date.

Converting z values to probabilities

A z value may be converted to the probability of not meeting the target date by using the graph in Fig. 5.5.

The advantages of PERT

We have seen that by requesting multivalued activity duration estimates and calculating expected dates, PERT focuses attention on the uncertainty of forecasting. We can use the technique to calculate the standard deviation for each task and use this to rank them according to their degree of risk. Using this ranking, we can see, for example, that activity F is the one over which we have greatest uncertainty, whereas activity C should, in principle, give us relatively little cause for concern.

If we use the expected times and standard deviations for forward passes through the network we can, for any event or activity completion, estimate the probability of meeting any set target. In particular, by setting target dates along the critical path, we can focus on those activities posing the greatest risk to the project's schedule.

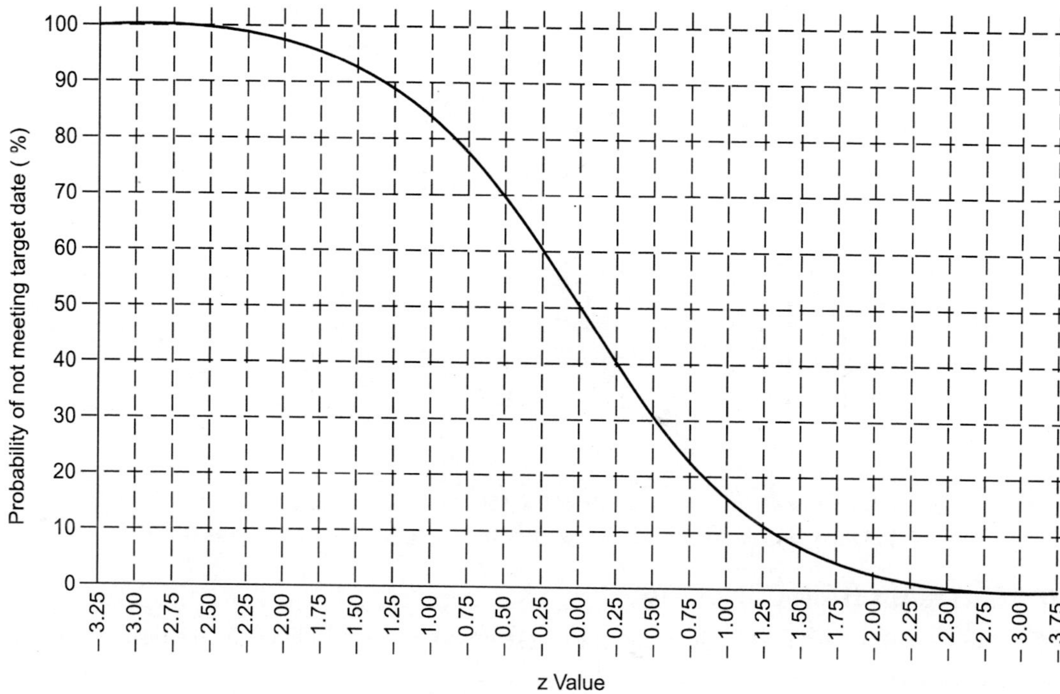

Fig. 5.5 The probability of obtaining a value within *z* standard deviations of the mean for a normal distribution.

5.9 MONTE CARLO SIMULATION

As an alternative to the PERT technique, and to provide a greater degree of flexibility in specifying likely activity durations, we can use Monte Carlo simulation techniques to evaluate the risks of not achieving deadlines. The basis of this technique involves calculating activity completion times for a project network a large number of times, each time selecting activity times randomly from a set of estimates for each activity. The results can then be tabulated, summarized or displayed as a graph such as that shwon in Fig. 5.6.

Activity duration estimates can be specified in a variety of forms, depending upon the information available. If, for example, we have historic data available about the durations of similar activities, we might be able to specify durations as a probability distribution. With less information available we should, at least, be able to provide three time estimates as used by PERT.

The calculation required for this is clearly extensive as we may have to carry out the forward pass through the network many hundreds of times before obtaining a representative selection of possible completion times. Fortunately, there are a number of packages available for carrying out Monte Carlo simulation. Some will exchange data with project-scheduling applications and some interface to standard spreadsheet software. The majority of these packages will apply Monte Carlo risk analysis to cost and resoruce as well as duration estimates.

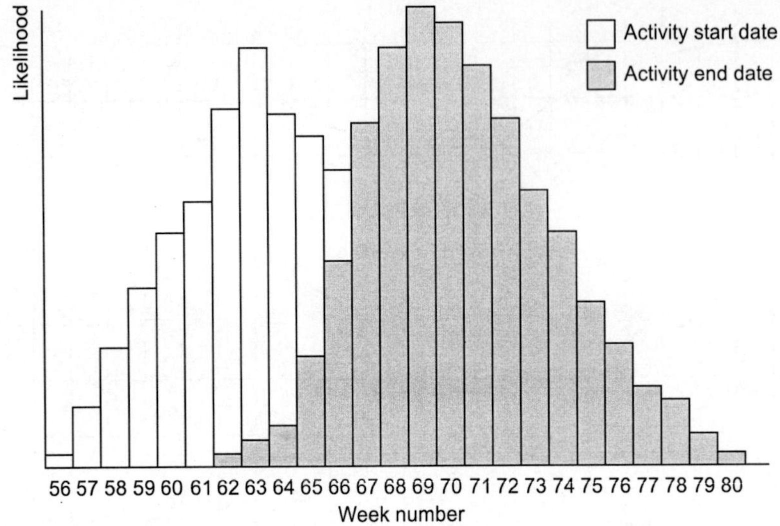

Fig. 5.6 Risk profile for an activity generated using Monte Carlo simulation

5.10 CRITICAL CHAIN CONCEPTS

The duration chosen as the target might be the one that seems to be the most likely. In the discussion above on the PERT risk technique this was one of the three numbers specified for each activity. When the project is actually being executed, the project manager will be forced to focus on the activities where the actual durations exceed the target. Activities which are actually completed before the target date are likely to be overlooked. These early completions, propoerly handled, could put some time in hand that might still allow the project to meet its target completion date if the later activities are delayed.

Fig. 5.7 shows the findings of Michiel van Genuchten, a researcher who analysed the reasons for delays in the completion of software development tasks. This bar chart shows that about 30% of activities were finished on time, while 9% were a week early and 17% were a week late. The big jump of 21 percentage points between being a week early and being on time is compatible with the 'Parkinson's Law' principle that '*work expands to fill the time available*'. This tendency should not be blamed on inherent laziness. Van Genuchten found that the most common reason for delay was the time that had to be spent on non-project work. It seems that developers used spare time provided by generous estimates to get on with other urgent work.

One approach which attempts to solve some fo these problems is the application of the *critical chain* concept originally developed by Eliyahu Goldratt. In order to demonstrate the principles of this approach, the example shown in Fig. 5.4 will be reworked as a Gantt chart. Fig. 5.8 shows what the Gantt chart for this project might look like if a 'traditional approach' were adopted and the pessimistic durations were used to supply 'comfort times' for each activity.

The general steps in the Critical Chain approach are as follows.

- The target date given to the developer is one where it is estimated that there is a 50% chance of success – this approximates to the expected time identified in the PERT risk method.
- Working backwards from the target completion date, each activity is shceduled to start as late as possible. Among other things, this should reduce the chance of staff being pulled off the project on to other work. It also makes every activity 'critical'. If it is larte the whole project is late.

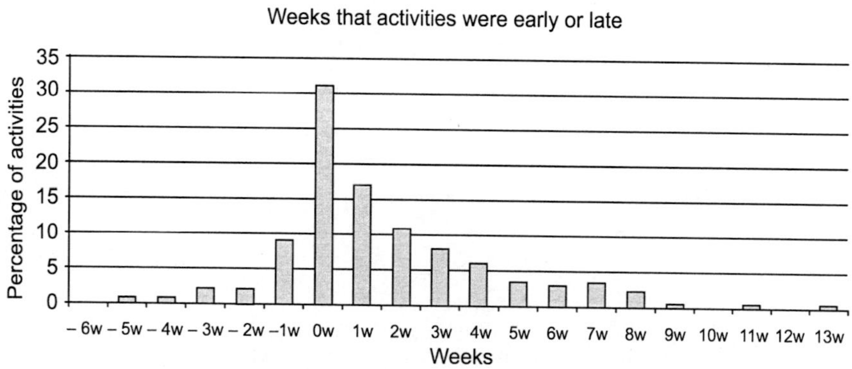

Fig. 5.7 Percentage of activities early of late (after Genuchen, 1991)

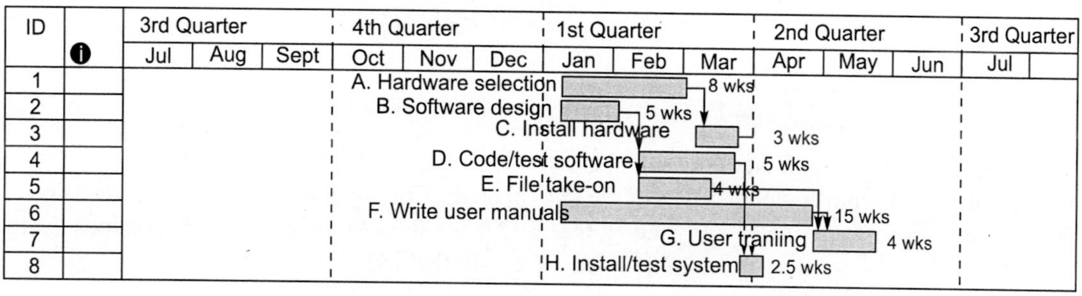

Fig. 5.8 Gantt chart – 'traditional' planning approach

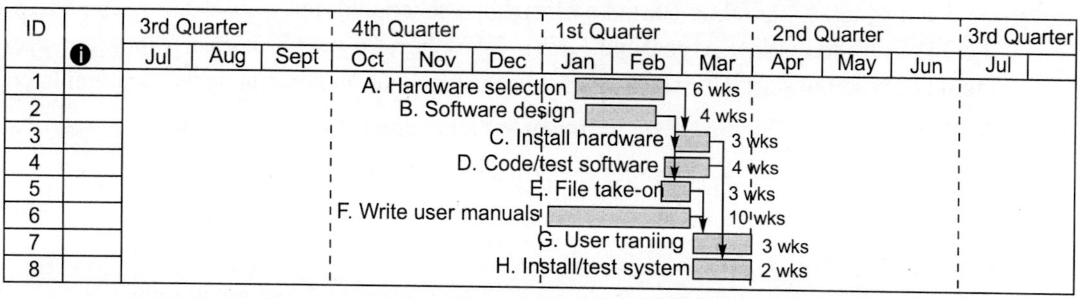

Fig. 5.9 Gantt chart – plan using 'expected times' and starting all activities as late as possible

In the example shown in Fig. 5.9, all the most likely durations have been used and eveyr activity is scheduled to start as late as possible as long as the project is completed by 1 March.

- To cope with activity overruns, a *project buffer* is inserted at the end of the project before the target completion date. This buffer is the equivalent of 50% of the length of the 'comfort zone' that has been removed from the *critical chain*. The critical chain is the longest chain of acitivities in the project, taking account of both task and resource dependencies. This is different from the *critical path* as the latter only takes account of task dependencies. A *resource dependency* is where one activity has to wait for a resource (usually a person) which is being used by another activity to become available. If an activity is late it will push the project completion date further into the project buffer.
- Buffers are also inserted into the project shedule where a subsidiary chain of activities feeds into the critical chain. These *feeding buffers* could be set at 50% of the length of the 'comfort zone' removed from the subsidiary or *feeding chain*.

Fig. 5.10 shows the results of this process. The critical chain in this example happens to be the same as the critical path, that is activities F and G which have comfort zones of 5 and 1 weeks respectively, making a total of 6 weeks. The project buffer is therefore 3 weeks.

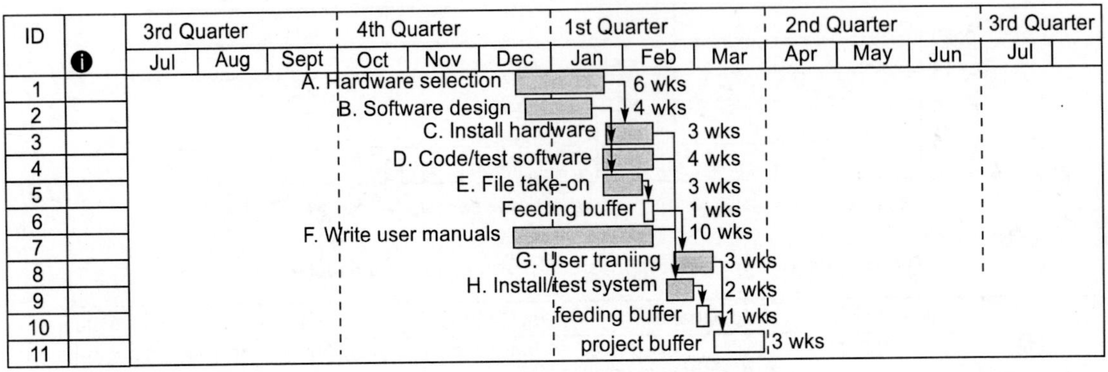

Fig. 5.10 Gantt chart - feeding buffers added

Subsidiary chains feed into the critical chain where activity H links into the project buffer and where activity E links into G which is part of the critical chain. For the first butter the duration would be 50% of the saved comfort zone of A, C and H, that is $(2 + 0 + 0.5)/2 = 1.25$ weeks. In the second case it would be 50% of the comfort zones of activities B and E, that is 1 week.

When the project is executed, the following principles are followed:

- No chain of tasks should be started earliear than scheduled, but once it has been started it should be finished as soon as possible – this invokes the *relay race principle*, where developers should be ready to start their tasks as soon as the previous, dependent, tasks are completed.
- Buffers are divided into three zones: green, amber and red, each of an even (33%) size;
- *green*, where no action is required if the project completion date creeps into this zone;
- *amber*, where an action plan is formulated if the project completion dates moves into this zone;
- *red*, where the action plan above is executed if the project penetrates this zone.

A Gantt chart is a horizontal bar chart developed as a production control tool in 1917 by Henry L. Gantt, an American engineer and social scientist. It is frequently used in project management.

A Gantt chart is useful for tracking and reporting progress, as well as for graphically displaying a schedule. Gantt chart are often used to report progress because they represent an easily understood picture of project status. However, Gantt charts are not an ideal tool for project control.

Gantt chart provides a graphical illustration of a schedule that helps to plan, coordinate, and track specific tasks in a project. Gantt charts may be simple versions created on graph paper or more complicated, automated versions created using project management tools.

Project management tools incorporating Gantt charts include PRINCE, MacProject, Microsoft Project and Microsoft Excel.

Purpose of a Gantt chart is to present a project schedule that shows the relationship of activities over time. Gantt charts are a project planning tool that can be used to represent the timing of tasks required to complete a project. Because Gantt charts are simple to understand and easy to construct, they are used by most project managers for all but the complex projects.

CONCEPTUAL SHORT QUESTIONS WITH ANSWERS

1. What are the different roles in SCM?

Ans. There are four major roles in SCM:

(*i*) Configuration Manager (*ii*) Change Control Board (CCB)

(*iii*) Developer (*iv*) Auditor

2. Give some of these risks, categorize them as either 'managerial' or as 'technical' risks.

(*a*) Behaviour of purchased software is incompatible with published standard.

(*b*) User not willing to use user interface as selected.

(*c*) The component is delivery date is later then planned.

Ans. (*a*) Technical Risk (*b*) Non-Technical Risk

(*c*) Management Risk.

3. How is risk reduction different from risk mitigation?

Ans. Risk reduction means reducing the likelihood of risk occuring. While risk mitigation is the action taken to reduce the impact of the risk. Mitigation is related with contigency planning *i.e.*, a planned action.

4. Are there any SCM standards for

(*a*) Military applications (*b*) Non-Military applications

Ans. (*a*) Military application software standards are:

MIL-STD-483, DOD-STD-480A,

MIL-STD-1521A.

(*b*) Non-Military or ANSI/IEEE standards are:

IEEE Std. NO. 828-1983,

Std. No. 1042-1987 and Std. No. 1028-1988.

REVIEW QUESTIONS

1. What is the difference between version control and change control?

2. How is risk reduction different from risk mitigation?

3. What is risk leverage (RL)? How will you compute it?

4. What is risk management and why it is important?

5. Discuss common sources and types of risks in software development projects and strategies to deal with them.

6. Differentiate between risk planning, risk avoidance, risk reduction and risk acceptance.

7. Discuss the common source and types of risks in software development projects and strategies to deal with them.

8. Suggest a contingency plan for the following risks:

(*a*) Project manager may leave in the middle of the project.

(*b*) Customer may aks for compression of delivery schedule.

9. Suggest a contingency plan for the following:

(*a*) Funds budget may fall short of 10% in the last phase of life cycle.

(*b*) System designer may have to leave for USA after 2 months.

(*c*) Software requires interface with other technologies on which the project team has no experience.

10. What is a Gantt charts?

CHAPTER 6

RESOURCE ALLOCATION

INSIDE THIS CHAPTER

6.1 INTRODUCTION

In this chapter we shall see how to match the activity plan to available resources and, where necessary, assess the efficacy of changing the plan to fit the resources, Fig. 6.1 shows where resource allocation is applied in Step Wise.

In general, the allocation of resources to activities will lead us to review and modify the ideal activity plan. It may cause us to revise stage or project completion dates. In any event, it is likely to lead to a narrowing of the time-spans within which activities may be scheduled.

The final result of resource allocation will normally be a number of schedules including:

- **activity schedule** indicating the planned start and completion dates for each activity;
- **resource schedule** showing the dates on which each resource will be required and the level of that requirement;
- **cost schedule** showing the planned cumulative expenditure incurred by the use of resources over time.

6.2 THE NATURE OF RESOURCES (*MDU 2010-11, 2012-13*) (*08 Marks*)

A resource is any item or person required for the execution of the project. This covers many things – from paperclips to key personnel – and it is unlikely that we would wish to itemize every resource required, let alone draw up a schedule for their use! Stationery and other standard office supplies, for example, need not normally he the concern of the project manager – ensuring there is always an adequate supply is the role of the office manager. The project manager must concentrate on those resources where there is a possibility that, without planning, they might not be sufficiently available when required.

Some resources, such as a project manager, will be required for the duration of the project whereas others, such as a specific software developer, might be required for a single activity. The former, while vital to the success of the project, does not require the same level of scheduling as the latter. Individual programmers, for example, might be committed to working on a number of projects and it will be important to book their time well in advance. In general, resources will fall into one of seven categories.

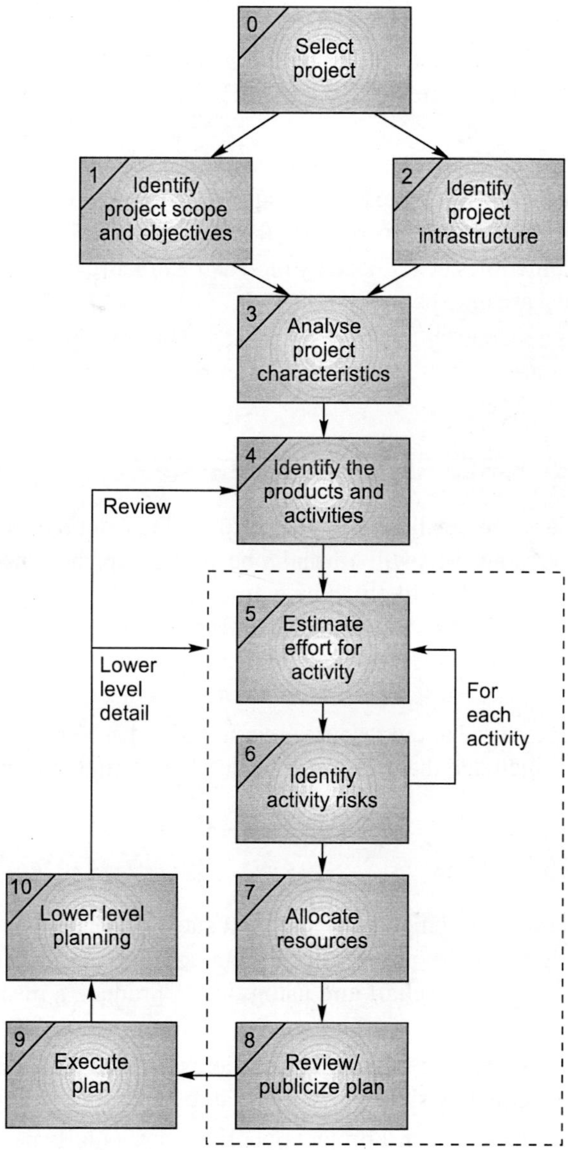

Fig. 6.1 Resource allocation is carried out as Step 7.

- **Labour.** The main items in this category will be members of the development project team such as the project manager, systems analysts and software developers. Equally important will be the quality assurance team and other support staff and any employees of the client organization who might be required to undertake or participate in specific activities.
- **Equipment.** Obvious items will include workstations and other computing and office equipment. We must not forget that staff also need basic equipment such as desks and chairs.
- **Materials.** Materials are items that are consumed, rather than equipment that is used. They are of little consequence in most software projects but can be important for some – software that is to be widely distributed might, for example, require supplies of floppy disks to be specially obtained.
- **Space.** For projects that are undertaken with existing staff, space is normally readily available. If any additional staff (recruited or contracted) should be needed then office space will need to be found.
- **Services.** Some projects will require procurement of specialist services – development of a wide area distributed system, for example, requires scheduling of telecommunications services.
- **Time.** Time is the resource that is being offset against the other primary resources – project time-scales can sometimes be reduced by increasing other resources and will almost certainly be extended if they are unexpectedly reduced.
- **Money.** Money is a secondary resource – it is used to buy other resources and will be consumed as other resources are used. It is similar to other resources in that it is available at a cost – in this case interest charges.

6.3 IDENTIFYING RESOURCE REQUIREMENTS

(MDU 2008-09) (05 Marks)

The first step in producing a resource allocation plan is to list the resources that will be required along with the expected level of demand. This will normally be done by considering each activity in turn and identifying the resources required. It is likely, however, that there will also be resources required that are not activity specific but are part of the project's infrastructure (such as the project manager) or required to support other resources (office space, for example, might be required to house contract software developers).

At this stage, it is necessary that the resource requirements list be as comprehensive as possible – it is better that something is included that may later be deleted as unnecessary than to omit something essential.

6.4 SCHEDULING RESOURCES

(MDU 2012-13, 2013-14) (10 Marks)

Having produced the resource requirements list, the next stage is to map this onto the activity plan to assess the distribution of resources required over the duration of the project. This is best done by representing the activity plan as a bar chart and using this to produce a resource histogram for each resource.

Each activity has been scheduled to start at its earliest start date – a sensible initial strategy, since we would, other things being equal, wish to save any float to allow for contingencies.

Changing the level of resources on a project over time, particularly personnel, generally adds to the cost of a project. Recruiting staff has costs and even where staff are transferred internally, time will be needed for familiarization with the new project environment.

An additional problem with an uneven resource histogram is that it is more likely to call for levels of resource beyond those available. Fig. 6.2 illustrates how, by adjusting the start date of some activities and splitting others, a resource histogram can, subject to constraints such as precedence requirements, be smoothed to contain resource demand at available levels. The different letters represent staff working on a series of module testing tasks, that is, one person working on task A, two on tasks B and C etc.

In Fig. 6.2, the original histogram was created by scheduling the activities at their earliest start dates. The resource histogram shows the typical peaked shape caused by earliest start date scheduling and calls for a total of nine staff where only five are available for the project.

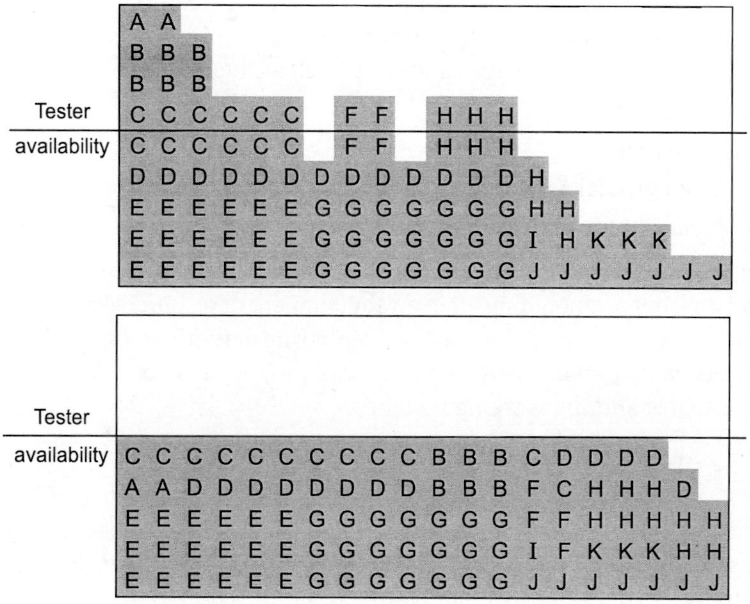

Fig. 6.2 A resource histogram showing demand for staff before and after smoothing.

By delaying the start of some of the activities, it has been possible to smooth the histogram and reduce the maximum level of demand for the resource. Notice that some activities, such as C and D, have been split. Where non-critical activities can be split they can provide a useful way of filling troughs in the demand for a resource, but in software projects it is difficult to split tasks without increasing the time they take.

Some of the activities call for more than one unit of the resource at a time – activity F, for example, requires two programmers, each working for two weeks. It might be possible to reschedule this activity to use one programmer over four weeks although that has not been considered in this case.

In practice, resources have to be allocated to a project on an activity-by-activity basis and finding the 'best' allocation can be time consuming and difficult. As soon as a member of the project team is allocated to an activity that activity acquires a scheduled start and finish date and the team member becomes unavailable for other activities for that period. Thus, allocating a resource to one activity limits the flexibility for resource allocation and scheduling of other activities.

It is therefore helpful to prioritize activities so that resources can be allocated to competing activities in some rational order. The priority must always be to allocate resources to critical path activities and then to those activities that are most likely to affect others. In that way, lower priority activities are made to fit around the more critical, already scheduled activities.

Of the various ways of prioritizing activities, two are described below:

- **Total float priority.** Activities are ordered according to their total float, those with the smallest total float having the highest priority. In the simplest application of this method, activities are allocated resources in ascending order of total float. However, as scheduling proceeds, activities will be delayed (if resources are not available at their earliest start dates) and total floats will be reduced. It is therefore desirable to recalculate floats (and hence reorder the list) each time an activity is delayed.
- **Ordered list priority.** With this method, activities that can proceed at the same time are ordered according to a set of simple criteria. An example of this is Burman's priority list, which takes into account activity duration as well as total float:
    1. shortest critical activity;
    2. critical activities;
    3. shortest non-critical activity;
    4. non-critical activity with least float;
    5. non-critical activities.

Unfortunately, resource smoothing, or even containment of resource demand to available levels, is not always possible within planned time-scales – deferring activities to smooth out resource peaks often puts back project completion. Where that is the case, we need to consider ways of increasing the available resource levels or altering working methods.

6.5 CREATING CRITICAL PATHS

Scheduling resources can create new critical paths. Delaying the start of an activity because of lack of resources will cause that activity to become critical if this uses up its float. Furthermore, a delay in completing one activity can delay the availability of a resource required for a later activity. If the later one is already critical then the earlier one might now have been made critical by linking their resources.

6.6 COUNTING THE COST

The discussion so far has concentrated on trying to complete the project by the earliest completion date with the minimum number of staff. We have seen that doing this places constraints on when activities can be carried out and increases the risk of not meeting target dates.

Alternatively, we could have considered using additional staff or lengthening the overall duration of the project. The additional costs of employing extra staff would need to be compared to the costs of delayed delivery and the increased risk of not meeting the scheduled date.

6.7 BEING SPECIFIC

Allocating resources and smoothing resource histograms is relatively straightforward where all resources of a given type can be considered more or less equivalent. When allocating labourers to activities in a large building project we need not distinguish among individuals – there are likely to be many labourers and they may be treated as equals so far as skills and productivity are concerned.

This is seldom the case with software projects. With the exception of extremely large projects it makes sense to allocate individual members of staff to activities as early as possible, as this can lead us to revise our estimate of their duration.

In allocating individuals to tasks, a number of factors need to be taken into account.

- **Availability.** We need to know whether a particular individual will be available when required. Reference to the departmental work plan determines this but the wise project manager will always investigate the risks that might be involved – earlier projects might, for example, over-run and affect the availability of an individual.
- **Criticality.** Allocation of more experienced personnel to activities on the critical path often helps in shortening project durations or at least reduces the risk of overrun.
- **Risk.** We saw how to undertake activity risk assessment in the previous chapter. Identifying those activities posing the greatest risk, and knowing the factors influencing them, helps to allocate staff. Allocating the most experienced staff to the highest risk activities is likely to have the greatest effect in reducing overall project uncertainties. More experienced staff are, however, usually more expensive.
- **Training.** It will benefit the organization if positive steps are taken to allocate junior staff to appropriate non-critical activities where there will be sufficient slack for them to train and develop skills. There can even be direct benefits to the particular project since some costs may be allocated to the training budget.
- **Team Building.** The selection of individuals must also take account of the final shape of the project team and the way they will work together.

6.8 PUBLISHING THE RESOURCE SCHEDULE

In allocating and scheduling resources we have used the activity plan (a precedence network in the case of the examples in this chapter), activity bar charts and resource histograms. Although good as planning took they are not the best way of publishing and communicating project schedules. For this we need some form of work plan. Work plans are commonly published either as lists or charts.

6.9 COST SCHEDULES

(MDU 2010-11) (08 Marks)

It is now time to produce a detailed cost schedule showing weekly or monthly costs over the life of the project. This will provide a more detailed and accurate estimate of costs and will serve as a plan against which project progress can be monitored.

Calculating cost is straightforward where the organization has standard cost figures for staff and other resources. Where this is not the case, then the project manager will have to calculate the costs.

In general, costs are categorized as follows:

- **Staff costs.** These will include staff salaries as well as the other direct costs of employment such as the employer's contribution to social security funds, pension scheme contributions, holiday pay and sickness benefit. These are commonly charged to projects at hourly rates based on weekly work records completed by staff. Note that contract staff are usually charged by the week or month – even when they are idle.
- **Overheads.** Overheads represent expenditure that an organization incurs, which cannot be directly related to individual projects or jobs including space rental, interest charges and the costs of service departments (such as personnel). Overhead costs can be recovered by making

a fixed charge on development departments (in which case they usually appear as a weekly or monthly charge for a project), or by an additional percentage charge on direct staff employment costs. These additional charges or oncosts can easily equal or exceed the direct employment costs.

- **Usage charges.** In some organizations projects are charged directly for use of resources such as computer time (rather than their cost being recovered as an overhead). This will normally be on an 'as used' basis.

6.10 THE SCHEDULING SEQUENCE

Going from an ideal activity plan to a costed schedule can be represented as a sequence of steps, rather like the classic waterfall life-cycle model. In the ideal world, we would start with the activity plan and use this as the basis for our risk assessment. The activity plan and risk assessment would provide the basis for our resource allocation and schedule from which we would produce cost schedules.

The interplay between the plans and schedules is complex – any change to any one will affect each of the others. Some factors can be directly compared in terms of money – the cost of hiring additional staff can be balanced against the costs of delating the project's end date. Some factors, however, are difficult to express in money terms (the cost of an increased risk, for example) and will include an element of subjectivity.

While good project planning software will assist greatly in demonstrating the consequences of change and keeping the planning synchronized, successful project scheduling is largely dependant upon the skill and experience of the project manager in juggling the many factors involved.

CONCEPTUAL SHORT QUESTIONS WITH ANSWERS

1. What is "Software Safety"?

Ans. Software Saftey is a Software Quality Assurance (SQA) activity. It focuses on the identification and assessment of potential hazards that may have negative impact on software and can fail the entire system.

2. Prove that the man-power utilization, $m(t)$, is given by the formula:

$$m(t) = \frac{dy}{dx} = 2k \text{ at } e^{-at^2}$$

where k is area under curve $[0, \infty]$ a is parameter affecting shape of curve t is elapsed time.

$\frac{dy}{dt}$ is man power utilization/time.

Ans.

$$m(t) = \frac{dy}{dt} = 2k \text{ at } e^{-at^2} \quad ...(i)$$

$$\int_0^{\infty} \frac{dy}{dt} = y(t) = k[1 - e^{-at^2}]$$

$$y(0) = 0$$

$$y(\infty) = k$$

To find relationship between 't_d' (peak time) and 'a', we find 2nd derivative *i.e.*,

$$\frac{d^2y}{dt^2} = 2k\, ae^{-at^2}\, [1 - 2\, at^2] = 0$$

or $$t_d^2 = \frac{1}{2a}$$

or $$a = \frac{1}{2t_d^2} \qquad \text{...(ii)}$$

Put (*ii*) in (*i*); we get

$$m(t) = \frac{2k}{2t_d^2} te^{-\frac{t^2}{2t_d^2}} = \frac{k}{t_d^2} te^{-\frac{t^2}{2t_d^2}} \qquad \text{...(iii)}$$

Now, at $t = t_d$, peak manning is obtained *i.e.*, $m(t_d) = m_0$ (say)
Putting $t = t_d$ in (*iii*), we get,

$$m(t_d) = \frac{k}{t_d^2} t_d e^{-\frac{t_d^2}{2t_d^2}} = \frac{k}{t_d} e^{-1/2}$$

$$m_0 = \frac{k}{t_d\sqrt{e}}$$

where k = Total Effort in person/years
m = No. of persons at peak
t_d = Delivery time in years

Now, consider once again,

$$m'(t) = \frac{d^2y}{dt^2} = 2k\, ae^{-at^2}\,(1 - 2at^2)$$

Now, for $$t = 0,\ m'(0) = 2ka = \frac{2k}{2t_d^2} = \frac{k}{t_d^2}$$

or $$m'(0) = \frac{k}{t_d^2} = \text{D} = \text{Difficulty (person years)}$$

But $$m_0 = \frac{k}{t_d\sqrt{e}} \text{ (just proved)}$$

$\therefore$ $$\text{D} = \frac{k}{t_d^2} = \frac{m_0\sqrt{e}}{t_d} \text{ proved.}$$

REVIEW QUESTIONS

1. What do you understand by the term resource allocation? Discuss its nature, requirements and scheduling.
2. Explain resource scheduling. How the proper resource scheduling can lead to effective cost cutting?
3. On a large project, it is often be the responsibility of a team leader to allocate tasks to individuals. Why might it be unsatisfactory to leave such allocations entirely to the discretion of the team leader?

CHAPTER 7

MONITORING AND CONTROL

INSIDE THIS CHAPTER

7.1 INTRODUCTION

Project monitoring and control is a key phase in the project management cycle. Properly executing this process will contribute significantly to successful project completion. It consists of all those processes which are performed to observe project execution so that potential problems can be identified in a timely manner and corrective actions can be taken when necessary to control the execution of the project.

Once work schedules have been published and the project is under way, attention must be focused on ensuring progress. This requires monitoring of what is happening, comparison of actual achievement against the schedule and, where necessary, revision of plans and schedules to bring the project as far as possible back on target.

In earlier chapters we have stressed the importance of producing plans that can be monitored. For example, ensuring that activities have clearly defined and visible completion points. We will discuss how information about project progress is gathered and what actions must be taken to ensure a project meets its targets.

The final part of this chapter discusses how we can deal with changes that are imposed from outside namely, changes in requirements.

In multi-phase projects, the monitoring and controlling process also provides feedback between project phases, in order to implement corrective or preventive actions to bring the project into compliance with the project management plan.

7.2 CREATING THE FRAMEWORK

(MDU 2010-11, 2012-13) (10 Marks)

Exercising control over a project and ensuring that targets are met is a matter of regular monitoring, finding out what is happening, and comparing it with current targets. If there is a mismatch between the planned outcomes and the actual ones then either replanning is needed to bring the project back on target or thc target will have to be revised. Fig. 7.1 illustrates a model of the project control cycle and shows how, once the initial project plan has been published, project control is a continual process of monitoring progress against that plan and, where necessary, revising the plan to take account of deviations. It also illustrates the important steps that must be taken after completion of the project so that the experienced gained in any one project can feed into the planning stages of future projects, thus allowing us to learn from past mistakes.

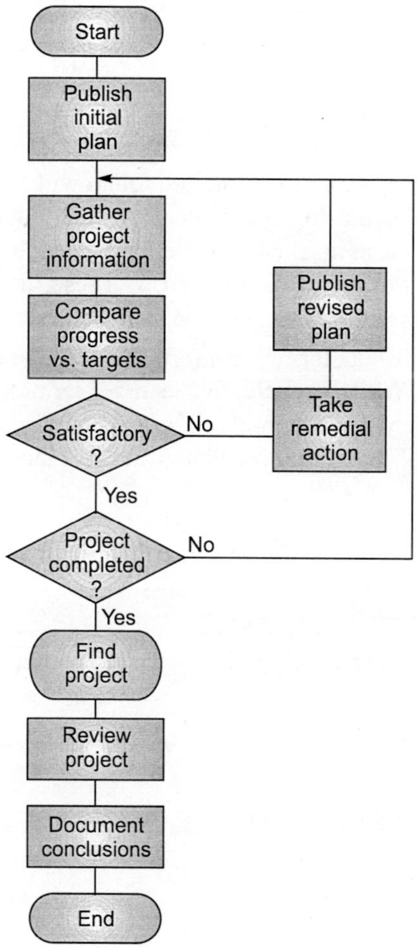

Fig. 7.1 The project control cycle.

In practice we are normally concerned with departures from the plan in four dimensions: delays in meeting target dates, shortfalls in quality, inadequate functionality, and costs going over target. In this chapter we are mainly concerned with the first and last of these.

Responsibility

The overall responsibility for ensuring satisfactory progress on a project is often the role of the *project steering committee* or *Project Board.* Day to day responsibility will rest with the project manager and, in all but the smallest of projects, aspects of this can be delegated to team leaders.

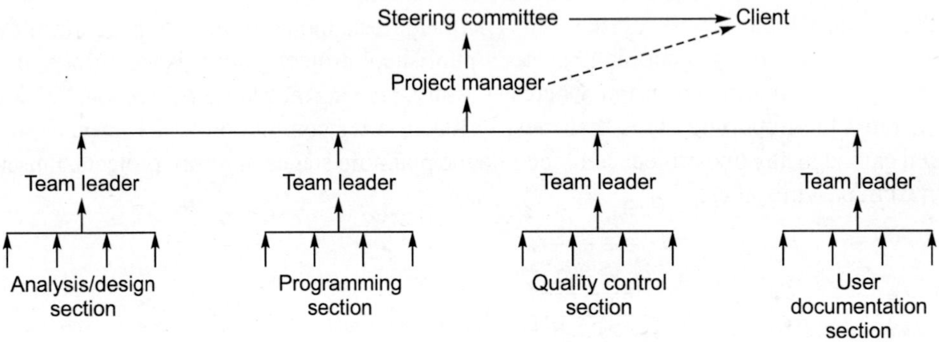

Fig. 7.2 Project reporting structures.

Fig. 7.2 illustrates the typical reporting structure found with medium and large projects. With small projects (employing around half a dozen or fewer staff) individual team members usually report directly to the project manager, but in most cases team leaders will collate reports on their section's progress and forward summaries to the project manager. These, in turn, will be incorporated into project-level reports for the steering committee and, via them or directly, progress reports for the client.

Reporting may be oral or written, formal or informal, or regular or ad hoc and some examples of each type are given in Table 7.1. While any effective team leader or project manager will be in touch with team members and available to discuss problems, any such informal reporting of project progress must be complemented by formal reporting procedures and it is those we are concerned with in this chapter.

Table 7.1: Categories of reporting

Report type	*Examples*	*Comment*
Oral formal regular	weekly or monthly progress meetings	while reports may be oral formal written minutes should be kept
Oral formal ad hoc	end-of-stage review meetings	while largely oral, likely to receive and generate written reports
Written formal regular	job sheets, progress reports	normally weekly using forms
Written formal ad hoc	exception reports, change reports	
Oral informal ad hoc	canteen discussion, social interaction	often provides early warning: must be backed up by formal reporting

Assessing progress

Progress assessment will normally be made on the basis of information collected and collated at regular intervals or when specific events occur. Wherever possible, this information will be objective and tangible whether or not a particular report has been delivered, for example. However, such end of activity deliverables might not occur sufficiently frequently throughout the life of the project. Here progress assessment will have to rely on the judgement of the team members who are carrying out the project activities.

Setting checkpoints

It is essential to set a series of checkpoints in the initial activity plan. Checkpoints may be:

- regular (monthly, for example);
- tied to specific events such as the production of a report or other deliverable.

Taking snap-shots

The frequency with which the manager needs to receive information about progress will depend upon the size and degree of risk of the project or that part of the project under their control. Team leaders, for example, need to assess progress daily (particularly when employing inexperienced staff) whereas project managers may find weekly; or monthly reporting appropriate. In general, the higher the level, the less frequent and less detailed the reporting needs to be.

There are, however, strong arguments in favour of formal weekly collection of information from staff carrying out activities. Collecting data at the end of each week ensures that information is provided while memories are still relatively fresh and provides a mechanism for individuals to review and reflect upon their progress during the past few days.

Major, or project level, progress reviews will generally take place at particular point during the life of a project commonly known as *review points* or *control points*. PRINCE 2, for example, designates a series of checkpoints where the status of work in a project or for a team is reviewed. At the end of each project Stage, PRINCE 2 provides for an End Stage Assessment where an assessment of the project and consideration of its future are undertaken.

7.3 COLLECTING THE DATA

As a rule, managers will try to break down long activities into more controllable tasks of one or two weeks duration. However, it will still be necessary to gather information about partially completed activities and, in particular, forecasts of how much work is left to be completed. It can be difficult to make such forecasts accurately.

Where there is a series of products, partial completion of activities is easier to estimate. Counting the number of record specifications or screen layouts produced, for example, can provide a reasonable measure of progress.

In some cases, intermediate products can be used as in activity milestones. The first successful compilation of a Cobol program, for example, might be considered a milestone even though it is not the final product of the activity code and test.

Partial completion reporting

Many organizations use standard accounting systems with weekly time sheets to charge staff time to individual jobs. The staff time booked to a project indicates the work carried out and the charges to the project. It does not, however, tell the project manager what has been produced or whether tasks are on schedule.

It is therefore common to adapt or enhance existing accounting data collection systems to meet the needs of project control. Weekly time sheets, for example, are frequently adapted by breaking jobs down to activity level and requiring information about work done in addition to time spent.

Asking for estimated completion times can be criticized on the grounds that frequent invitations to reconsider completion dates deflects attention away from the importance of the originally scheduled targets and can generate an ethos that it is acceptable for completion dates to slip.

Risk reporting

One popular way of overcoming the objections to partial completion reporting is to avoid asking for estimated completion dates, but to ask instead for the team members' estimates of the likelihood of meeting the planned target date.

One way of doing this is the traffic light method. This consists of the following steps:

- identify the key (first level) elements for assessment in a piece of work;
- break these key elements into constituent elements (second level);
- assess each of the second level elements on the scale *green* for 'on target', *amber* for 'not on target but recoverable', and *red* for 'not on target and recoverable only with difficulty';
- review all the second level assessments to arrive at first level assessments;
- review first and second level assessments to produce an overall assessment.

7.4 VISUALIZING PROGRESS

(MDU 2012-13, 2013-14) (10 Marks)

Having collected data about project progress, a manager needs some way of presenting that data to greatest effect. In this section, we look at some methods of presenting a picture of the project and its future. Some of these methods (such as Gantt charts) provide a static picture, a single snap-shot, whereas others (such as time-line charts) try to show how the project has progressed and changed through time.

The Gantt chart

One of the simplest and oldest techniques for tracking project progress is the Gantt chart. This is essentially an activity bar chart indicating scheduled activity dates and durations frequently augmented with activity floats. Reported progress is recorded on the chart (normally by shading activity bars) and a 'today cursor' provides an immediate visual indication of which activities are ahead or behind schedule.

The slip chart

A slip chart is a very similar alternative favoured by some project managers who believe it provides a more striking visual indication of those activities that are not progressing to schedule—the more the slip line bends, the greater the variation from the plan. Additional slip lines are added at intervals and, as they build up, the project manager will gain an idea as to whether the project is improving (subsequent slip lines bend less) or not. A very jagged slip line indicates a need for rescheduling.

Ball charts

A somewhat more striking way of showing whether or not targets have been met is to use a ball chart. In this version of the ball chart, the circles indicate start and completion points for activities. The circles initially contain the original scheduled dates. Whenever revisions are produced these are added as second dates in the appropriate circle until an activity is actually started or completed when the relevant date replaces the revised estimate. Circles will therefore contain only two dates, the original and most recent target dates, or the original and actual dates.

Such charts are frequently placed in a prominent position and the colour coded balls provide a constant reminder to the project team. Where more than one team is working in close proximity, such a highly visible record of achievement can encourage competitiveness between teams.

Another advantage of ball charts over Gantt and slip charts is that they are relatively easy to keep upto date, only the dates and possibly colours need to be changed, whereas the others need to be redrawn each time target dates are revised.

The timeline

One disadvantage of the charts described so far is that they do not show clearly the slippage of the project completion date through the life of the project. Knowing the current state of a project helps in revising plans to bring it back on target, but analysing and understanding trends helps to avoid slippage in future projects.

The timeline chart is a method of recording and displaying the way in which targets have changed throughout the duration of the project.

The timeline chart is useful both during the execution of a project and as part of the post implementation review. Analysis of the timeline chart, and the reasons for the changes, can indicate failures in the estimation process or other errors that might, with that knowledge, be avoided in future.

7.5 COST MONITORING

Expenditure monitoring is an important component of project control. Not only in itself, but also because it provides an indication of the effort that has gone into (or at least been charged to) a project. A project might be on time but only because more money has been spent on activities than originally budgeted. A cumulative expenditure chart such as that shown in Fig. 7.3 provides a simple method of comparing actual and planned expenditure. By itself it is not particularly meaningful. Fig. 7.3 could, for example, illustrate a project that is running late or one that is one time but has shown substantial cots savings! We need to take account of the current status of the project activities before attempting to interpret the meaning of recorded expenditure.

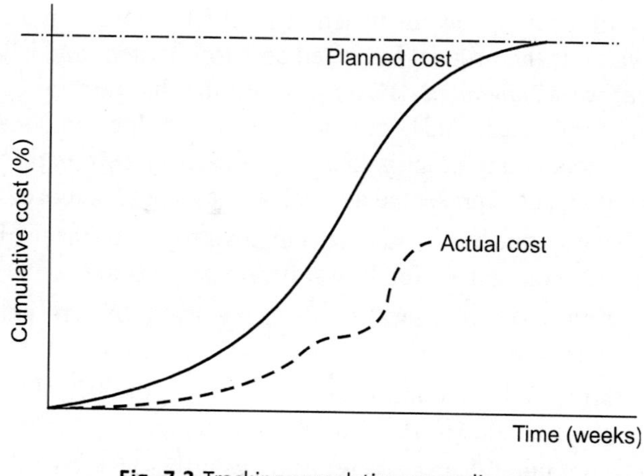

Fig. 7.3 Tracking cumulative expenditure.

Cost charts become much more useful if we add projected future costs calculated by adding the estimated costs of uncompleted work to the costs already incurred. Where a computer-based planning tool is used, revision of cost schedules is generally provided automatically once actual expenditure has been recorded. Fig. 7.4 illustrates the additional information available once the revised cost schedule is included—in this case it is apparent that the project behind schedule and over budget.

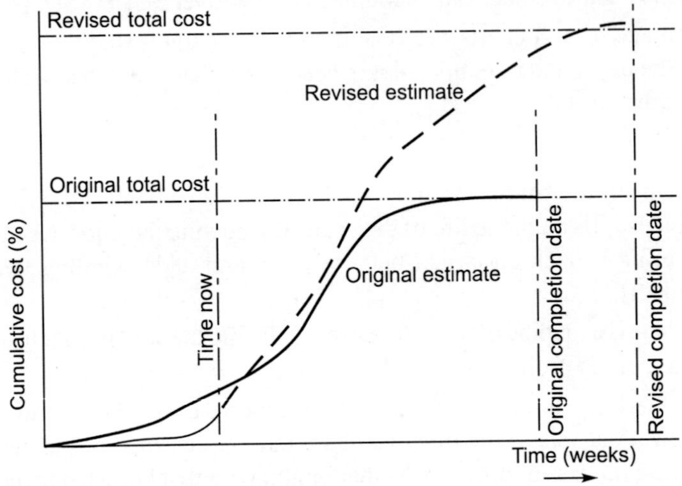

Fig. 7.4 The cumulative expenditure chart can also show revised estimates of cost and completion date.

7.6 EARNED VALUE *(MDU 2007-08, 2010-11, 2012-13, 2013-14)* (*10 Marks*)

Earned Value Management (EVM) technique is used to track the progress and status of a Project and Forecasts the liekly future performance of the project. It integrates the scope, schedule and cost of a project. This technique can be used to show past performance of the project, current performance of the project and predict the future performance of the project by use of statistical technique.

Earned Value Analysis has gained in popularity in recent years and may be seen as a refinement of the cost monitoring discussed in the previous section. Earned Value Analysis is based on assigning a 'value' to each task or work package (as identified in the WBS) based on the original expenditure forecasts. The assigned value is the original budgeted cost for the item and is known as the *baseline budget* or *budgeted cost of work scheduled* (BCWS). A task that has not started is assigned the value zero and when it has been completed, it, and hence the project, is credited with the value of the task. The total value credited to a project at any point is known as the *earned value* or *budgeted cost of work performed* (BCWP) and this can be represented as a value or as a percentage of the BCWS.

Where tasks have been started but are not yet complete, some consistent method of assigning an earned value must be applied. Common methods in software projects are:

- **The 0/100 technique:** Where a task is assigned a value of zero until such time that it is completed when it is given a value of 100% of the budgeted value;
- **The 50/50 technique:** Where a task is assigned a value of 50% of its value as soon as it is started and then given a value of 100% once it is complete;
- **The milestone technique:** Where a task is given a value based on the achievement of milestones that have been assigned values as part of the original budget plan.

Of these, we prefer the 0/100 technique. The 50/50 technique can give a false sense of security by over-valuing the reporting of activity starts. The milestone technique might be appropriate for activities with a long duration estimate but, in such cases, it is better to break that activity into a number of smaller ones.

The baseline budget

The first stage in setting up an earned value analysis is to create the *baseline budget.* The baseline budget is based on the project plan and shows the forecast growth in earned value through time. Earned value may be measured in monetary values but, in the case of staff-intensive projects such as software development, it is common to measure earned value in person-hours or workdays.

Monitoring earned value

Having created the baseline budget, the next task is to monitor earned value as the project progresses. This is done by monitoring the completion of tasks (or activity starts and milestone achievements in the case of the other crediting techniques).

As well as recording BCWP, the actual cost of each task can be collected as *actual cost of work performed*, ACWP. This is shown in Fig. 7.5 which, in this case, records the values as percentages of the total budget cost. Fig. 7.5 also illustrates the following performance statistics, which can be shown directly or derived from the earned value chart.

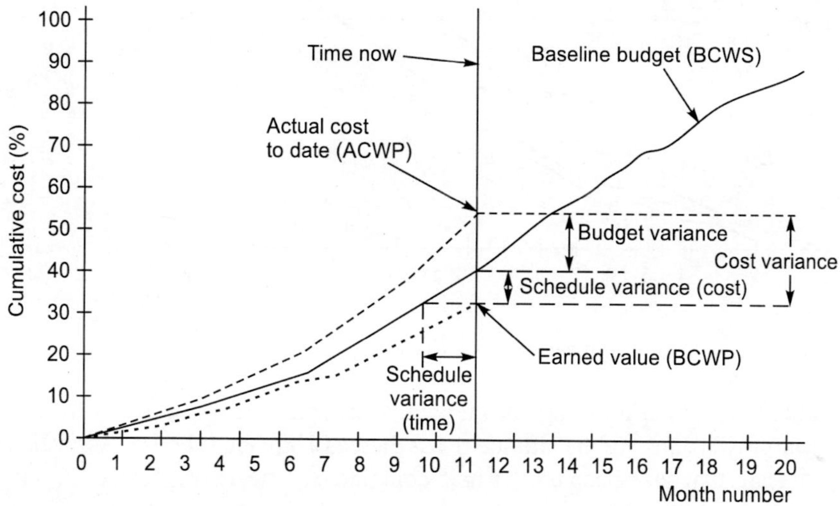

Fig. 7.5 An earned value tracking chart.

Budget variance. This can be calculated as ACWP-BCWS and indicates the degree to which actual costs differ from those planned.

Schedule variance. The schedule variance is measured in cost terms as BCWP-BCWS and indicates the degree to which the value of completed work differs from that planned.

Cost variance. This is calculated as BCWP-ACWP and indicates the difference between the budgeted cost and the actual cost of completed work. It is also an indicator of the accuracy of the original cost estimates.

Performance ratios. Two ratios are commonly tracked: the *cost performance index* (CPT-BCWP/ACWP) and the *schedule performance index* (SPI = BCWP/BCWS). They can be thought of as a 'value-for-money' indices. A value greater than one indicates that work is being completed better than planned whereas a value of less than one means that work is costing more than and/or proceeding more slowly than planned.

In the same way that the expenditure analysis in Fig. 7.3 was augmented to show revised expenditure forecasts, we can argument the simple Earned Value tracking chart with forecasts as illustrated in Fig. 7.6.

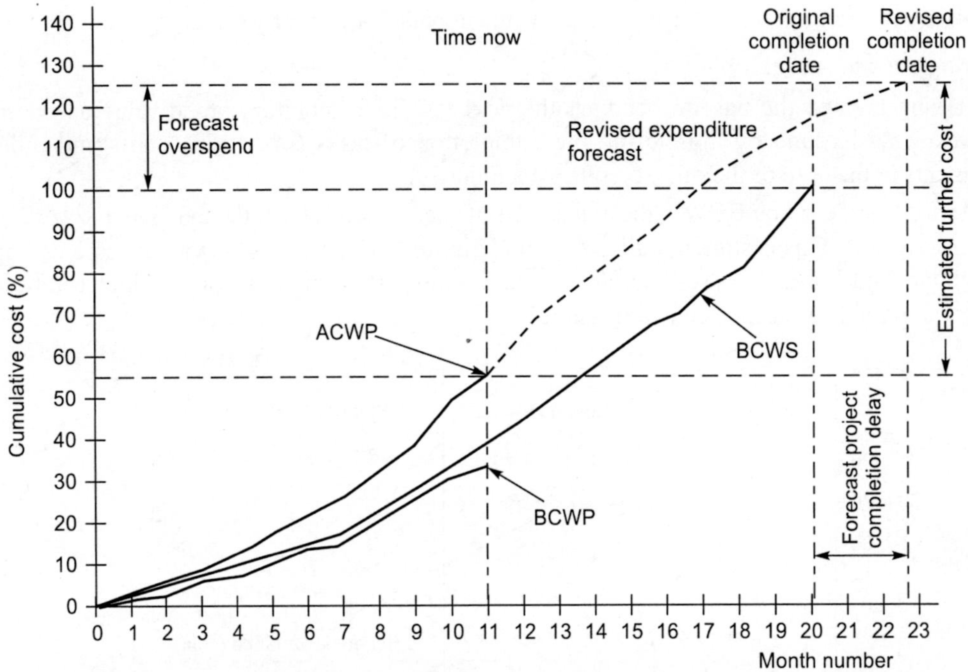

Fig. 7.6 An earned value chart with revised forecasts.

Earned value analysis has not yet gained universal acceptance for use with software development projects, perhaps largely because of the attitude that, whereas a half-built house has a value reflected by the labour and materials that have been used, a half-completed software project has virtually no value at all. This is to misunderstand the purpose of earned value analysis, which, as we have seen, is a method for tracking what has been achieved on a project—measured in terms of the budgeted costs of completed tasks or products.

7.7 PRIORITIZING MONITORING

So far we have assumed that all aspects of a project will receive equal treatment in terms of the degree of monitoring applied. We must not forget, however, that monitoring takes time and uses resources that might sometimes be put to better use.

In this section we list the priorities we might apply in deciding levels of monitoring.

- **Critical path activities.** Any delay in an activity on the critical path will cause a delay in the completion date for the project. Critical path activities are therefore likely to have a very high priority for close monitoring.

- **Activities with no free float.** A delay in any activity with no free float will delay at least some subsequent activities even though, if the delay is less than the total float, it might not delay the project completion date. These subsequent delays can have serious effects on our resource schedule as a delay in a subsequent activity could mean that the resources for that activity will become unavailable before that activity is completed because they are committed elsewhere.
- **Activities with less than a specified float.** If any activity has very little float it might use up this float before the regular activity monitoring brings the problem to the project manager's attention. It is common practice to monitor closely those activities with less than, say, one week free float.
- **High risk activities.** A set of high risk activities should have been identified as part of the initial risk profiling exercise. If we are using the PERT three-estimate approach we will designate as high risk those activities that have a high estimated duration variance. These activities will be given close attention because they are most likely to overrun or overspend.
- **Activities using critical resources.** Activities can be critical because they are very expensive (as in the case of specialized contract programmers). Staff or other resources might be available only for a limited period, especially if they are controlled outside the project team. In any event, an activity that demands a critical resource requires a high level of monitoring.

7.8 GETTING THE PROJECT BACK TO TARGET

Almost any project will, at one time or another, be subject to delays and unexpected events. One of the tasks of the project manager is to recognize when this is happening (or, if possible, about to happen) and, with the minimum delay and disruption to the project team, attempt to mitigate the effects of the problem. In most cases, the project manager tries to ensure that the scheduled project end date remains unaffected. This can be done by shortening remaining activity durations or shortening the overall duration of the remaining project in the ways described in the next section.

It should be remembered, however, that this might not always be the most appropriate response to disruptions to a plan. There is little point in spending considerable sums in overtime payments in order to speed up a project if the customer is not overly concerned with the delivery date and there is no other valuable work for the team members once this project is completed.

There are two main strategies to consider when drawing up plans to bring a project back on target—shortening the critical path or altering the activity precedence requirements.

Shorten the critical path

The overall duration of a project is determined by the current critical path, so speeding up non-critical path activities will not bring forward a project completion date.

Extolling staff to 'work harder' might have some effect, although frequently a more positive form of action is required, such as increasing the resources available for some critical activity. Fact-finding, for example, might be speeded up by allocating an additional analyst to interviewing users. It is unlikely, however, that the coding of a small module would be shortened by allocating an additional programmer—indeed, it might be counterproductive because of the additional time needed organizing and allocating tasks and communicating.

Resource levels can be increased by making them available for longer. Thus, staff might be asked to work overtime for the duration of an activity and computing resources might be made available at times (such as evenings and week-ends) when they might otherwise be inaccessible.

Where these do not provide a sufficient solution, the project manager might consider allocating more efficient resources to activities on the critical path or swapping resources between critical and non-critical activities. This will be particularly appropriate with staff—an experienced programmer should be significantly more productive than a more junior member of the team.

By such means we can attempt to shorten the timescale for critical activities until such time as either we have brought the project back to schedule or further efforts prove unproductive or not cost-effective. Remember, however, that shortening a critical path often causes some other path, or paths, to become critical.

Reconsider the precedence requirements

If attempting to shorten critical activities proves insufficient, the next step is to consider the constraints by which some activities have to he deferred pending completion of others. The original project network would most probably have been drawn up assuming 'ideal' conditions and 'normal' working practices. It might be that, to avoid the project delivering late, it is now worth questioning whether as yet unstarted activities really do have to await the completion of others. It might, in a particular organization, be 'normal' to complete system testing before commencing user training. In order to avoid late completion of a project it might, however, be considered acceptable to alter 'normal' practice and start training earlier.

One way to overcome precedence constraints is to subdivide an activity into a component that can start immediately and one that is still constrained as before. For example, a user handbook can be drawn up in a draft form from the system specification and then be revised later to take account of subsequent changes.

If we do decide to alter the precedence requirements in such a way, it is clearly important to be aware that quality might be compromised and to make a considered decision to compromise quality where needed. It is equally important to assess the degree to which changes in work practices increase risk. It is possible, for example, to start coding a module before its design has been completed. It would normally, however, be considered foolhardy to do so since, as well as compromising quality, it would increase the risk of having to redo some of the coding once the final design had been completed and thus delay the project even further.

7.9 CHANGE CONTROL *(MDU 2008-09, 2012-13, 2013-14) (10 Marks)*

So far in this chapter, we have assumed that the nature of the tasks to be carried out has not changed. Careful control of these changes is needed because an alteration in one document often implies changes to other documents and the system products based on that document.

Configuration librarian's role

Control of changes and documentation ought to be the responsibility of someone who may variously be named the Configuration Librarian, the Configuration Manager or Project Librarian. Among this person's duties would be:

- the identification of all items that are subject to change control;
- the establishment and maintenance of a central repository of the master copies of all project documentation and software products;

- the setting up and running of a formal set of procedures to deal with changes;
- the maintenance of records of who has access to which library items and the status of each library item (*e.g.* whether under development, under test or released).

Change control procedures

A simple change control procedure for operational systems might have the following steps:

1. One or more users might perceive a need for a modification to a system and ask for a change request to be passed to the development staff.

2. The user management consider the change request and if they approve it pass it to the development management.

3. The development management delegate a member of staff to look at the request and to report on the practicality and cost of carrying out the change. They would, as part of this, assess the products that would be affected by the change.

4. The development management report back to the user management on the findings and the user management decide whether, in view of the cost quoted, they wish to go ahead.

5. One or more developers are authorized to take copies of the master products that are to be modified.

6. The copies are modified. In the case of software components this would involve modifying the code and recompiling and testing it.

7. When the development of new versions of the product has been completed the user management will be notified and copies of the software will be released for user acceptance testing.

8. When the user is satisfied that the products are adequate they will authorize their operational release. The master copies of configuration items will be replaced.

Changes in scope of a system

A common occurrence with IS development product is for the size of the system gradually to increase. One cause of this is changes to requirements that are requested by users.

The scope of a project needs to be carefully monitored and controlled. One way is to re-estimate the system size in terms of SLOC or function points at key milestones.

CONCEPTUAL SHORT QUESTIONS WITH ANSWERS

1. How delayed projects can be brought back?

Ans. Delayed projects can be brought back by

(*i*) Shortening activity times on critical path

(*ii*) Relaxing some precedence constraints.

2. How project monitoring and control can be done?

Ans. Project monitoring and control can be done by measuring maturity of a project.

3. What are the two main strategies for getting the project back to target?

Ans. The two main strategies are:

(*i*) Shortening the critical path

(*ii*) Giving priority to precedenced requirements.

4. What do you understand by negative value of Cost Variance (CV)?

Ans. A negative value of Cost Variance (CV) means the project is undercost.

5. What do you understand by negative value of Schedule Variance (SV)?

Ans. A negative value of Schedule Variance (SV) means the project is behind schedule?

REVIEW QUESTIONS

1. Describe EVA.
2. Write a short note on EVA.
3. What are the two main strategies for getting the project back to target?
4. Suggest some approaches or methods to shorten the project duration time.
5. How can you show the different levels of planning with respect to WBS, PERT/CPM methodologies?

CHAPTER 8

MANAGING CONTRACTS AND PEOPLE

INSIDE THIS CHAPTER

8.1 INTRODUCTION

The buying in of both goods and services, rather than 'doing it yourself', is attractive when money is available but other, less flexible, types of resource, especially staff time, are in short supply. However, there are hazards for organizations who adopt this policy. Many of these potential dangers arise from the fact that considerable staff time and attention will still be needed to manage a contracted out project successfully.

Also, it needs hardly be said that potential suppliers are more likely to be flexible and accommodating before any contract has been signed than they will be afterwards – especially if the contract is for a fixed price. All this points to the need for as much forethought and planning with an acquisition project as with an internal development project.

In the remainder of this chapter, we will first discuss the different types of contract that can be negotiated. We will then follow through the general steps that ought to be followed when placing a contract. The issues that ought to be considered when drafting a contract are then examined. We conclude by describing some of the things that need to be done while the contract is actually being executed.

Note that the bargaining position of the customer will be much stronger if their business is going to be very valuable. If you are buying a cut-price computer game from a local store, you are unlikely to be able to negotiate variations on the supplier's standard contract of sale! (Indeed, because of the inequality of the parties in such circumstances, such sales are subject to special consumer protection laws). It is reasonable for potential suppliers to weigh up carefully the time and money they are willing to spend responding to a customer's initial request, as there is no guarantee of their obtaining the final contract.

8.2 TYPES OF CONTRACT *(Important)*

The external resources required could be in the form of *services*. A simple example of this could be using temporary staff on short term contracts to carry out some project tasks.

A more far-reaching use of external services would be for the contractor not only to supply the new system but to also operate it on the customer's behalf.

On the other hand, the contract could be placed for the supply of a *completed software application*.

This could be:

- a *bespoke* system, that is, a system that is created from scratch specifically for one customer;
- *off-the-shelf,* which you buy 'as is' – this is sometimes referred to as *shrink-wrapped* software;
- *customized off-the-shelf* (COTS) software – this is a basic core system, which is modified to meet the needs of a particular customer.

Where equipment is being supplied then, in English law, this may be regarded as a contract for the supply of *goods*. In the case of the supply of software this may be regarded as supplying a service (to write the software) or the granting of a *licence* (or permission) to use the software, which remains in the ownership of the supplier. These distinctions will have legal implications.

Another way of classifying contracts is by the way that the payment to suppliers is calculated. We will look at:

- fixed price contracts;
- time and materials contracts;
- fixed price per delivered unit contracts.

Fixed price contracts

As the name implies, in this situation a price is fixed when the contract is signed. The customer knows that, if there are no changes in the contract terms, this is the price to be paid on the completion of the work. In order for this to be effective, the customer's requirement has to be known and fixed at the outset. In other words, when the contract is to construct a software system, the detailed requirements analysis must already have been carried out. Once the development is under way, the customer will not be able to change their requirements without renegotiating the price of the contract.

The advantages of this method are the following:

- **Known customer expenditure.** If there are few subsequent changes to the original requirements, then the customer will have a known outlay.
- **Supplier motivation.** The supplier has a motivation to manage the delivery of the system in a cost-effective manner.

The disadvantages include the following:

- **Higher prices to allow for contingency.** The supplier absorbs the risk for any errors in the original estimate of product size. To reduce the impact of this risk, the supplier will add a margin when calculating the price to be quoted in a tender.
- **Difficulties in modifying requirements.** The need to change the scope of the requirements sometimes becomes apparent as the system is developed – this can cause friction between the supplier and customer.

- **Upward pressure on the cost of changes.** When competing against other potential suppliers, the supplier will try and quote as low a price as possible. If, once the contract is signed, further requirements are put forward, the supplier is in a strong position to demand a high price for these changes.
- **Threat to system quality.** The need to meet a fixed price can mean that the quality of the software suffers.

Time and materials contracts

With this type of contract, the customer is charged at a fixed rate per unit of effort, for example, per staff-hour. At the start of the project, the supplier normally provides an estimate of the overall cost based on their current understanding of the customer's requirements, but this is not the basis for the final payment.

The advantages of this approach are the following:

- **Ease of changing requirements.** Changes to requirements are dealt with easily. Where a project has a research orientation and the direction of the project changes as options are explored, then this can be an appropriate method of calculating payment.
- **Lack of price pressure.** The lack of price pressure can allow better quality software to be produced.

The disadvantages of this approach are:

- **Customer liability.** The customer absorbs all the risks associated with poorly defined or changing requirements.
- **Lack of incentives for supplier.** The supplier has no incentive to work in a cost-effective manner or to control the scope of the system to be delivered.

Because the supplier appears to be given a blank cheque, this approach does not normally find favour with customers. However, the employment of contract development staff, in effect, involves this type of contract.

Fixed price per unit delivered contracts

This is often associated with function point (FP) counting. The size of the system to be delivered is calculated or estimated at the outset of the project. The size of the system to be delivered might be estimated in lines of code, but FPs can be more easily and reliably derived from requirements documents. A price per unit is also quoted. The final price is then the unit price multiplied by the number of units. Table 8.1 shows a typical schedule of prices.

Table 8.1: A schedule of charges per function point

Function point count	*Function design cost per FP*	*Implementation cost per FP*	*Total cost per FP*
Upto 2,000	$242	$725	$967
2,001–2,500	$255	$764	$1,019
2,501–3,000	$265	$793	$1,058
3,001–3,500	$274	$820	$1,094
3,501–4,000	$284	$850	$1,134

The company who produced this table, RDI Technologies of the USA, in fact charge a higher fee per FP for larger systems. For example, a system to be implemented contains 2,600 FPs. The overall charge would be 2,000 × \$967, plus 500 × \$1,1019, plus 100 × \$1,058.

One problem that has already been noted is that the scope of the system to be delivered can grow during development. The software supplier might first carry out the system design. From this design, an FP count could be derived. A charge could then be made for design work based on the figures in the 'Function design cost per FP' column. This, if the designed system was counted at 1,000 FPs, would be 1000 × \$242 = \$242,000. If the design were implemented, and the actual software constructed and delivered, then the additional 1000 × \$725 = \$725,000 would be charged. If the scope of the system grows because the users find new requirements, these new requirements would be charged at the combined rate for design and implementation. For example, if new requirements amounting to 100 extra FPs were found, then the charge for this extra work would be \$967 × 100 = \$96,700.

The advantages of this approach are as follows:

- **Customer understanding.** The customer can see how the price is calculated and how it will vary with changed requirements.
- **Comparability.** Pricing schedules can be compared.
- **Emerging functionality.** The supplier does not bear the risk of increasing functionality.
- **Supplier efficiency.** The supplier still has an incentive to deliver the required functionality in a cost-effective manner (unlike with time and materials contracts).
- **Life cycle range.** The requirements do not have to be definitively specified at the outset. Thus the development contract can cover both the analysis and design stages of the project.

The disadvantages of this approach are as follows:

- **Difficulties with software size measurement.** Lines of code can easily be inflated by adopting a verbose coding style. With FPs, there can be disagreements about what the FP count should really be: in some cases, FP counting rules might be seen as unfairly favouring either the supplier or customer. Users, in particular, will almost certainly not be familiar with the concept of FPs and special training might be needed for them. The solution to these problems might be to employ an independent FP counter.
- **Changing requirements.** Some requested changes might affect existing transactions drastically but not add to the overall FP count. A decision has to be taken about how to deal with these changes. A change made late in the development cycle will almost certainly require more effort to implement than one made earlier.

To reduce the last difficulty, one suggestion from Australia has been to vary the charge depending on the point at which they have been requested – see Table 8.2.

Table 8.2: Examples of additional charges for changed functionality

	Pre-acceptance testing handover	*Post-acceptance testing handover*
Additional FPs	100%	100%
Changed FPs	130%	150%
Deleted FPs	25%	50%

In addition to the three payment methods above, there are other options and permutations of options. For instance, the implementation of an agreed specification could be at a fixed price, with provision for any additions or changes to the requirements to be charged for on a per FP basis. Another example could be where the contractor has to buy in large amounts of equipment, the price of which might fluctuate through no fault of the contractor. In this case it is possible to negotiate a contract where the final price contains a fixed portion for labour plus an amount that depends on the actual cost of a particular component used.

An underlying problem with software can once again be seen when the questions of contractual obligations and payment are considered – namely its relative invisibility and its flexibility. These mean that system size and consequently development effort tends to be very difficult to judge. If contractors are realistically to quote firm prices for work, then the tasks they are asked to undertake must be carefully constrained. For example, it would be unrealistic for a contractor to be asked to quote a single price for all the stages of a development project: how can they estimate the construction effort needed when the requirements are not yet established? For this reason it will often be necessary to negotiate a series of contracts, each covering a different part of the system development life cycle.

Another way of categorizing contracts, at least initially, is according to the approach that is used in contractor selection:

- open;
- restricted;
- negotiated.

Open tendering process

In this case, any supplier can bid to supply the goods and services. Furthermore all bids that are compliant with the original conditions laid down in the *invitation to tender* must be considered and evaluated in the same way as all the others. With a major project where there are lots of bids and the evaluation process is time-consuming, this can be an expensive way of doing things.

In recent years, there has been a global movement towards removing artificial barriers that hamper businesses in one country supplying goods and services in another. Examples of this are the drives by bodies such GATT and the European Union to ensure that national governments and public bodies do not limit the granting of contracts to their fellow nationals without good reason. One element of this is the laying down of rules about how tendering processes should be carried out. In certain circumstances, these demand that an open tendering process be adopted.

Restricted tendering process

In this case, there are bids only from suppliers who have been invited by the customer. Unlike the open tendering process, the customer may at any point reduce the number of potential suppliers being considered. This is usually the best approach to be adopted. However, it is not without risk: where the resulting contract is at a fixed price, the customer assumes responsibility for the correctness and completeness of the requirements specified to the prospective suppliers. Defects in this requirements documentation sometimes allow a successful bidder subsequently to claim for additional payments.

Negotiated procedure

There are often, however, some good reasons why the restricted tendering process might not be the most suitable in some particular sets of circumstances. Say, for example, that there is a fire that destroys part of an office, including IT equipment. The key concern here is to get replacement equipment up and running as quickly as possible and there is no time to embark on a lengthy tendering process.

Another situation might be where a new software application had been successfully built and implemented by an outsider, but the customer decides to have some extensions to the system. As the original supplier has staff who have complete familiarity with the existing system, it might once again be inconvenient to approach other potential suppliers via a full tendering process.

In these cases, an approach to a single supplier might be justified. However, it takes little imagination to realise that approaching a single supplier can open the customer up to charges of favouritism and should be done only where there is a clear justification.

8.3 STAGES IN CONTRACT PLACEMENT *(MDU 2010-11, 2012-13)* *(08 Marks)*

We are now going to discuss the typical stages in awarding a contract.

Requirements analysis

Before potential supplier can be approached, you need to have a clear set of requirements. Two points need to be emphasized here. The first is that it is easy for this step to be skimped where the user has many day-to-day pressures and not much time to think about future developments. In this situation, it can be useful to bring in an external consultant to draw up a requirements document. Even here, users and their managers need to look carefully at the resulting requirements document to ensure that it accurately reflect their needs. As David Bainbridge has pointed out; '*the lack of, or defects in, the specification are probably the heart of most disputes resulting from the acquisition of computer equipment and software*'.

The requirements document might typically have sections with the headings shown in Table 8.3. This requirements document is sometimes called an operational requirement or OR.

Table 8.3: Main sections in a requirements document

Section name
1. Introduction
2. A description of any existing systems and the current environment
3. The customer's future strategy or plans
4. System requirements • mandatory • desirable
5. Deadlines
6. Additional information required from potential suppliers

The requirements define carefully the *functions* that need to be carried out by the new application and all the necessary *inputs* and *outputs* for these functions. The requirements should also state any *standards* with which there should be compliance, and the existing systems with which the new system needs to be compatible. As well as these functional requirements, there will also need to be operational and quality requirements concerning such matters as the required response times, reliability, usability and maintainability of the new system.

In general, the requirements document should state *needs* as accurately as possible and should avoid technical specifications of possible solutions. The onus should be placed on the potential suppliers to identify the technical solutions that they believe will meet the customer's stated needs. After all, they are the technical experts who should have access to the most up-to-date information about current technology.

Each requirement needs to be identified as being either *mandatory* or *desirable.*

- **Mandatory.** If a proposal does not meet this requirement then the proposal is to be immediately rejected. No further evaluation would be required.
- **Desirable.** A proposal might be deficient in this respect, but other features of the proposal could compensate for it.

Among the other details that should be included in the requirements document to be issued to potential suppliers would be requests for any information needed to help us judge the standing of organization itself. This could include financial reports, references from past customers and the CVs of key development staff.

Evaluation plan ***(Very Important)***

Having drawn up a list of requirements, we now need to draw up a plan of how the proposals that are submitted are to be evaluated. The situation will be slightly different if the contract is for a system that is to be specially written as opposed to an off-the-shelf application. In the latter case, it is the system itself that is being evaluated while in the former situation it is a proposal for a system.

First, a means of checking that all the mandatory requirements have been met needs to be identified. The next consideration is of how the desirable requirements can be evaluated. The problem here is weighing the value of one quality against another. The ISO 9126 standard, which is discussed in the chapter on software quality, can be used to decide that one system has more 'quality' than another, but if there is a difference in price between the two, we need to be able to estimate if the increase in quality is worth the additional price. Hence 'value for money' is often the key criterion.

Could a financial value be placed on this? If we were to cost clerical effort at £20 an hour and we knew that four hours of clerical effort a month went into entering staffing costs into the accounting computer system, then we could conclude that over a four year period (£20 an hour × 4 hours a month × 48 months), or £3840, would be saved. If system A has this feature and costs £1000 more than system B, which does not, then this would seem to give system A an advantage. If, however, system A cost £5000 more than B then the picture would be different.

It needs to be stressed that the costs to be taken into account are those for the whole of the lifetime of the proposed system, not just the costs of acquiring the system. Also, where the relationship with the supplier is likely to be ongoing, the supplier organization needs to be assessed as well as its products.

Invitation to tender

Having produced the requirements and the evaluation plan, it is now possible to issue the invitation to tender to prospective suppliers. Essentially, this will be the requirement document with a supporting letter, which may have additional information about how responses to the invitation are to be lodged. A deadline will be specified and it is hoped that by then a number of proposals with price quotations will have been received.

In English law, for a contract to exist there must be an offer on one side that must be accepted by the other side. The invitation to tender is not an offer itself but an invitation for prospective suppliers to make an offer.

Certain new problems now emerge. The requirements that have been laid down might be capable of being satisfied in a number of different ways. The customer needs to know not only a potential supplier's price but also the way in which they intend to satisfy the requirements – this will be particularly important where the contract is to build a new system from scratch.

In some relatively straight-forward cases, it would be enough to have some post-tender clarification and negotiation to resolve issues in the supplier's proposal. With more complex projects a more elaborate approach may be needed. One way of getting the detail of the suppliers' proposals elaborated is to have a two stage tendering process.

In the first stage, technical proposals are requested from potential suppliers who do not necessarily quote any prices at this stage. Some of these proposals can be dismissed out of hand as not being able to meet the mandatory requirements. With the remaining ones, discussions may be held with representatives of the suppliers in order to clarify and validate the technical proposals. The suppliers may be asked to demonstrate certain aspects of their proposals. Where shortcomings in the proposal are detected, the supplier may be given the opportunity to remedy these.

The result of these discussions could be a *Memorandum of Agreement* (MoA) with each prospective supplier. This is an acceptance by the customer that the proposed solution (which might have been modified during discussions) offered by the supplier satisfactorily meets the customer's requirement.

In the second stage, tenders are invited from all the suppliers who have signed individual Memoranda of Agreement. The tender would incorporate the MoA and would be concerned with the financial terms of a potential contract.

If a design has to be produced as part of the proposal made by a supplier in response to an invitation to tender, then a difficulty would be that the supplier would have to do a considerable amount of detailed design work with only a limited prospect of being paid for it. One way of reducing this burden is for the customer to choose a small number of likely candidates who will be paid a fee to produce design proposals. These can then be compared and the final contract for construction awarded to the most attractive proposal.

The ISO 12207 has a rather different approach. Once a contract for software construction is signed, the supplier develops a design, which then has to be agreed by the customer.

Evaluation of proposals

This needs to be done in a methodical and planned manner. We have already mentioned the need to produce an evaluation plan, which will describe how each proposal will be checked to see if it meets each requirement. This reduces risks of requirements being missed and ensures that all proposals are treated consistently. Otherwise, there is a risk that a proposal might be unfairly favoured because of the presence of a feature that was not requested in the original requirement.

It will be recalled that an application could be either bespoke, off-the-shelf, or customized. In the case of off-the-shelf packages, it would be the software itself that would be evaluated and it might be possible to combine some of the evaluation with acceptance testing. With bespoke development it would be a proposal that is evaluated, while COTS may involve elements of both. Thus different planned approaches would be needed in each case.

The process of evaluation may include:

- scrutiny of the proposal documents;
- interviewing suppliers' representatives;
- demonstrations;

- site visits;
- practical tests.

The proposal documents provided by the suppliers can be scrutinized to see if they contain features satisfying all the original requirements. Clarification might need to be sought over certain points. Any factual statements made by a supplier imply a legal commitment on their part if it influences the customer to offer the contract to that supplier. It is therefore important to get a written, agreed, record of these clarifications. The customer may take the initiative here by keeping notes of meetings and then writing afterwards to the suppliers to get them to confirm the accuracy of the notes. A supplier might, in the final contract document, attempt to exclude any commitment to any representations that have been made in pre-contract negotiations – the terms of the contract need to be scrutinized in this respect.

Where the delivered product is to be based on an existing product it might be possible to see a demonstration. A danger with demonstrations is that they can be controlled by the supplier and as a passive observer it is often difficult to maintain full attention for more than, say, half-an-hour. Because of this, the customer should produce a schedule of what needs to be demonstrated, ensuring that all the important features are seen in operation.

With off-the-shelf software, it should be possible to have actual access to the application. For example, a demonstration version, which closes itself down after 30 days, might be available. Once again a test plan is needed to ensure that all the important features are evaluated in a complete and consistent manner. Once a particular package emerges as the most likely candidate, it needs to be carefully investigated to see if there are any previously unforeseen factors that might invalidate this choice.

A frequent problem is that while an existing application works well on one platform with a certain level of transactions, it does not work satisfactorily on the platform that the customer has or at the level of throughput that it would be subjected to in the customer's work environment. Demonstrations will not generally reveal this problem. Visits to operational sites already using the system will be more informative in this respect. In the last resort a special volume test could be conducted.

Eventually a decision will be made to award the contract to one of the suppliers. One of the central reasons for using a structured and, as far as possible, objective approach to evaluation is to be able to demonstrate that the decision has been made impartially and on merit. In most large organizations, placing a contract involves the participation of a second party within the organization, such as a contracts department, who can check that the correct procedures have been carried out. Also the final legal format of a contract will almost certainly require some legal expertise.

In any case, not only should the successful candidate be notified but the unsuccessful candidates should also be told of the decision. This is not simply a matter of courtesy: under GATT or EU rules, there is a legal requirement to do this in certain circumstances. It makes dealing with unsuccessful bidders easier if they can be given clear and objective reasons why their proposals did not find favour.

8.4 TYPICAL TERMS OF A CONTRACT *(MDU 2010-11, 2012-13, 2013-14)* *(20 Marks)*

Definitions

The terminology used in the contract document may need to be defined, for example, who is meant by the words 'client' and 'supplier'.

Form of agreement

For example, is it a contact of sale, a lease, or a licence? Also, can the subject of the contract, such as a licence to use a software application, be transferred to another party?

Goods and services to be supplied

Equipment and software to be supplied. This includes an actual list of the individual pieces of equipment to be delivered, complete with the specific model numbers.

Services to be provided. This covers such things as:

- training;
- documentation;
- installation;
- conversion of existing files;
- maintenance agreements;
- transitional insurance arrangements.

Ownership of the software

Who has ownership of the software? There are two key issues here: firstly, whether the customer can sell the software to others and, secondly, whether the supplier can sell the software to others. Where off-the-shelf software is concerned, the supplier often simply grants a license for you to use the software. Where the software is being written specially for a customer, then that customer will normally wish to ensure exclusive use of the software – they may object to software which they hoped would give them a competitive edge being sold to their rivals. They could do this by acquiring the copyright to the software outright or by specifying that they should have *exclusive use* of the software. This would need to be written into the contract. Where a core system has been customized by a supplier, then there is less scope for the customer to insist on exclusive use.

Where software is written by an employee as part of a contract of employment, it is assumed that the copyright belongs to the employer. Where the customer organization has contracted an external supplier to write software, the contract needs to make clear who is going to retain the copyright – it cannot, in this case, be automatically assumed it is the customer. The customer might have decided to take over responsibility for maintenance and further development once the software is delivered and in this case will need access to the source code. In other cases, where the customer does not have an adequate in-house maintenance function, the supplier can retain the source code, and the customer will have to approach the supplier for any further changes. There are many potential dangers with this, not the least being that the supplier could go out of business. An escrow agreement can be included in the contract so that up-to-date copies of the source code are deposited with a third party. In the United Kingdom, the National Computing Centre provide an escrow service.

Environment

Where physical equipment is to be installed, the demarcation line between the supplier's and customer's responsibilities with regard to such matters as accommodation and electrical supply needs to be specified. Where software is being supplied, the compatibility of the software with the existing hardware and operating system platforms would need to be confirmed.

Customer commitments

Even when work is carried out by external contractors, a development project still needs the participation of the customer. The customer will have to provide accommodation for the suppliers and perhaps other facilities such as telephone lines.

Acceptance procedures

Good practice would be to accept a delivered system only after it has undergone user acceptance tests. This part of the contract would specify such details as the time that the customer will have to conduct the tests, deliverables upon which the acceptance tests depend and the procedure for signing off the testing as completed.

Standards

This covers the standards with which the goods and services should comply. For example, a customer can require the supplier to conform to the ISO 12207 standard relating to the software life cycle and its documentation (or, more likely, a customized sub-set of the standard). Within the European Union, government customers with contracts for projects above a certain threshold value must, by law, ensure that the work conforms to certain standards.

Project and quality management ***(Important)***

The arrangements for the management of the project must be agreed. Among these would be frequency and nature of progress meetings and the progress information to be supplied to the customer. The contract could require that appropriate ISO 9000-series standards be followed. The ISO 12207 standard provides for the customer to have access to quality documentation generated internally by the supplier, so that the customer can ensure that there is adherence to standards.

Timetable

This provides a schedule of when the key parts of the project should he completed. This timetable will commit both the supplier and the customer. For example, the supplier might be able to install the software on the agreed date only if the customer makes the hardware platform available at that point.

Price and payment method

Obviously the price is very important! What also needs to be agreed is when the payments are to be made. The supplier's desire to be able to meet costs as they are incurred needs to be balanced by the customer's requirement to ensure that goods and services are satisfactory before parting with their money.

Miscellaneous legal requirements

This is the legal small print. Contracts often have clauses that deal with such matters the legal jurisdiction that will apply to the contract, what conditions would apply to the sub-contracting of the work, liability for damage to third parties, and liquidated damages. *Liquidated damages* are estimates of the financial losses that the customer would suffer if the supplier were to fall short of their obligations. It is worth noting that under English law, the penalties laid down in penalty clauses must reflect the actual losses the customer would suffer and cannot be unrealistic and merely punitive. Even this limitation will not be enough in some cases as far as the supplier is concerned. As computer systems assume increasingly critical roles in many organizations and in safety-critical systems can even be life-threatening in the case of malfunction, the possible consequential damage could be very great. Suppliers will not unnaturally try to limit this kind of liability. The courts (in England and Wales) have tended to look critically at such attempts at limiting liability, so that suppliers will, in the case of major contracts, take out insurance to cover such liabilities.

If there is a dispute, resorting to litigation, while being lucrative to the lawyers involved, is both time-consuming and expensive. An alternative is to agree that disputes be settled by *arbitration.* This requires that any dispute be referred to an expert third party whose decision as to the facts of the case is binding. Even this procedure is seldom quick and inexpensive and another option is *alternative dispute resolution* where a third party acts as a mediator who has only an advisory capacity and attempts to broker an agreement between the two sides.

8.5 CONTRACT MANAGEMENT (*MDU 2010-11, 2011-12, 2012-13*) (*10 Marks*)

We now need to consider the communications between the supplier and the customer while the work contracted for is being carried out. It would probably suit all concerned if the contractor could be left to get on with the work undisturbed. However, at certain *decision points,* the customer needs to examine work already done and make decisions about the future direction of the project. The project will require representatives of the supplier and customer to interact at many points in the development cycle – for example, users need to be available to provide information needed to carry out effective detailed interface design.

This interaction, or other external factors, often leads to changes being needed, which effectively vary the terms of the contract and so a careful change control procedure is needed. Each of these topics will now be tackled in a little more detail.

When a contract is being negotiated, certain key points in the project can be identified where customer approval is needed before the project can proceed. For example, a project to develop a large system can be divided into increments. For each increment there could be an interface design phase, and the customer needs to approve the designs before the increment is built. There could also be a decision point between increments.

For each decision point, the deliverables to be presented by the suppliers, the decisions to be made by the customer and the outputs from the decision point all need to be defined. These decision points have added significance if payments to the supplier are based on them. Not only the supplier but also the customer has responsibilities with respect to these decision points – for example, the supplier should not be unnecessarily delayed while awaiting customer approval of some interim deliverable.

Where work is contracted out there will be a general concern about the quality of that work. The ISO 12207 standard envisages the possibility of there being agents, employed independently of the supplier or customer, who will carry out verification, validation and quality assurance. It also allows for joint reviews of project processes and products, the nature of which needs to be clearly agreed when the contract is negotiated, otherwise the supplier might claim unwarranted interference in their work.

As the system is developed a need to change certain of the requirements often emerges. As noted earlier, essentially, this is varying the terms of the contract. Oral evidence is not normally admissible to contradict, add to, or vary the terms of a written contract, so that agreed changes need to be documented properly. An effective change control procedure is therefore needed to record requests for changes, along with the supplier's agreement to them and any fees for the additional work.

It could happen that the supplier does not meet one or more of their legal obligations. This might be through no fault of theirs, if, for example, the customer has caused the delay by being tardy in giving the necessary approvals for intermediate products. If no action is taken when the default occurs, this can be taken to imply that the customer in fact condones the failure and this can lead to the loss of a right to legal redress. The customer should therefore protect their legal rights by officially notifying the supplier as soon as possible that the failure has been recognized. It will be recalled that under English law any claim for liquidated damages should be based on actual losses. From the point where the default occurs, the customer needs to keep an accurate record of the actual losses incurred as a result of the default including any consequential losses.

8.6 ACCEPTANCE

(Important)

When the work has been completed, the customer needs to take action to carry out acceptance testing. The contract might put a time limit on how long acceptance testing can take, so the customer must be organized to carry out this testing before the time limit for requesting corrections expires.

We have already noted that some software houses are rather cursory with their pre-acceptance testing: the implication seeming to be that they would rather the users spent their time on testing than they themselves. This imposition can be reduced by asking to approve the supplier's internal test plans. An associated pitfall is that once the main development work is completed, the supplier not unnaturally wants to reallocate the most productive staff to other projects. The customer can find that all their problem reports are being dealt with by relative junior members of the supplier's staff, who might not be familiar with all aspects of the delivered system.

Part or all of the payment to the supplier will depend on this acceptance testing. Sometimes part of the final payment will be retained for a period of operational running and is eventually paid over if the levels of reliability are as contracted for. There is usually a period of warranty during which the supplier should fix any errors found for no charge. The supplier might suggest a very short warranty period of say 30 days. It is in the customer's interests to negotiate a more realistic period of say at least 120 days.

8.7 INTRODUCTION TO MANAGING PEOPLE AND ORGANIZING TEAMS

(MDU 2009-10, 2012-13, 2013-14) (10 Marks)

First, we will look at some aspects of organizational behaviour (OB) research. There will be three concerns: staff selection, staff development and, which will be dealt with in more detail, staff motivation.

We will look at how the project leader can encourage effective group working and decision making while balancing this, where needed, by purposeful leadership. The final part of the chapter looks at some of the more formal aspects of organizational structures.

The issues raised in this chapter have impacts at all stages of project planning and execution but in particular at the following points (see also Fig. 8.1):

- although perhaps having little control over organizational structure, the project leader needs to be aware of its implications (Step 2):
- the scope and nature of activities can be set in a way that will enhance staff motivation (Step 4);
- many risks to project success relate to staffing (Step 6);
- the qualities of individual members of staff should be taken into account when allocating staff to activities (Step 7).

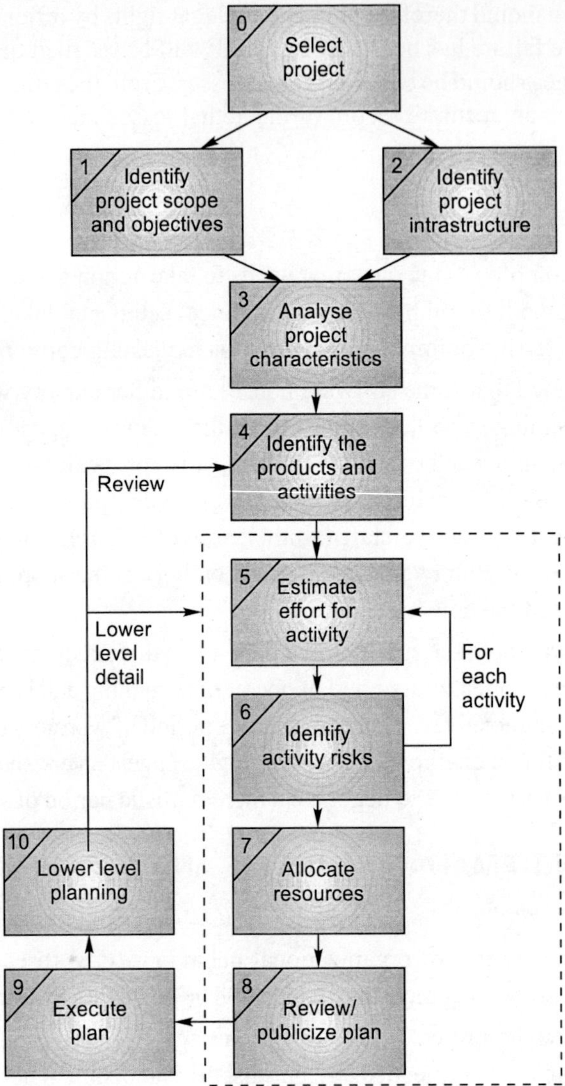

Fig. 8.1 Some places in the Step Wise framework where staffing concerns are important.

8.8 UNDERSTANDING BEHAVIOUR

People with practical experience of working on projects invariably identify the handling of people as one of the most important aspects of project management.

The field of social science known as organizational behaviour (OB) helps. This has evolved theories that try to explain people's behaviour and that tend to be structured 'If A is the situation then B is likely to result'. Attempts are then made to observe behaviour or to conduct experiments where variables for A and B are measured and a statistical relationship between the two variables is sought. Unlike physical science, it is rarely, if ever, the case that it can be said that B must always follow A.

A major problem is that in the real world there is bound to be a very wide range of influences on a situation, many of which will not be apparent to the observer. It is therefore difficult to decide which set of research findings is relevant. A danger is that we end up with a set of maxims that are little better than superstitions. However, it is to hoped that by examining these questions people can become more sensitive and thoughtful about the problems involved.

8.9 ORGANIZATIONAL BEHAVIOUR: A BACKGROUND *(MDU 2010-11) (08 Marks)*

The roots of studies in OB can be traced back to work done in the late 19th and early 20th centuries by Frederick Taylor. By studying the way that manual workers did tasks, he attempted to work out the most productive way of doing these tasks. The workers were then trained to do the work in this way.

Taylor had three basic objectives:

- to select the best person for the job;
- to instruct such people in the best methods;
- to give incentives in the form of higher wages to the best workers.

'Taylorism' is often represented as crude and mechanistic these days. Interestingly though, the Taylorist approach is one that is adopted, in part, in modern sports coaching. A coach will attempt to get a javelin thrower, for example, to throw a javelin in a very exact manner in order to get the maximum effect. In the more mundane world of software development, the growth of structured methods is an example of this emphasis on best practice.

During the 1920s, OB researchers discovered, while carrying out a now famous set of tests on the conditions under which staff worked best, that not only did a group of workers for whom conditions were improved increase their work-rates, but a control group, for whom conditions were unchanged, also increased their work-rates. Simply showing a concern for what workers did increased productivity. This illustrated how the state of mind of workers influenced their productivity.

The cash-oriented view of work of some managers can thus be contrasted with a more rounded vision of people in their place of work. The two attitudes were labelled *Theory X* and *Theory Y* by Donald McGregor.

Theory X holds that:

- the average human has an innate dislike of work;
- there is a need therefore for coercion, direction and control;
- people tend to avoid responsibility.

Theory Y, on the other hand, holds that:

- work is as natural as rest or play;
- external control and coercion are not the only ways of bringing about effort directed towards the company's ends;
- commitment to objectives is a function of the rewards associated with their achievement;
- the average human can learn to accept and further seek responsibility;
- the capacity to exercise imagination and other creative qualities is widely distributed.

One way of judging whether a manager espouses Theory X or Theory Y is to observe how the manager's staff react when the boss is absent: if there is no discernible change, then this is a Theory Y environment; if everyone visibly relaxes, it is a Theory X environment. McGregor's distinction between the two theories also draws attention to the way that expectations influence behaviour. If a manager (or teacher) assumes that you are going to work diligently and do well, then you are likely to try and meet these expectations.

8.10 SELECTING THE RIGHT PERSON FOR THE JOB *(MDU 2008-09, 2012-13) (10 Marks)*

Taylor stressed the need for the right person for the job. Many factors, such as the use of software tools and methodologies, affect programming productivity. However, one of the biggest differences in software development performance is among individuals. As early as 1968, a comparison of experienced professional programmers working on the same programming task found a ratio, in one case, of 1:25 between the shortest and longest time to code the program and, more significantly perhaps, of 1:28 for the time taken to debug it.

What sort of characteristics should they be looking for? Should they go, for example, for the experienced programmer or the new graduate with the first class mathematics degree? It is extremely dangerous to generalize but looking specifically at behavioural characteristics, the American researcher Cheney found that the most important influence on programmer productivity seemed to be experience. Mathematical aptitude had quite a weak influence in comparison.

The American researchers Couger and Zawacki found that computing people would appear to have much weaker 'social needs' than people in other professions. They quote Gerald Weinberg: *'If asked, most programmers probably say they prefer to work alone where they wouldn't be disturbed by other people.'* This is reflected in the problem that people attracted to writing software, and are good at it, will not make good managers later in their careers.

The recruitment process

Although this is an important matter, it has to be stressed that often project leaders have little choice about the people who will make up the teams – they have to make do with the 'materials that are to hand'. Recruitment might very well be regarded as an organizational responsibility: you might be recruiting someone who will, over a period of time, work in many different parts of the same organization.

Meredith Belbin usefully distinguishes between *eligible* and *suitable* candidates. An eligible candidate is one whose CV (curriculum vitae or resumed) shows, for example, the 'right' number of years in some previous post and the 'right' paper qualifications. Suitable candidates are those who can actually do the job well. An easy mistake is to select an eligible candidate who is not in fact suitable. Suitable candidates who are not technically eligible can, on the other hand, be ideal candidates because, once in post, they are more likely to remain loyal to the organization. Belbin suggests that selection methods that centre on the assessment of actual skills rather than past experience and also a willingness to provide training to make good minor gaps in expertise can be a more effective way of placing suitable staff. It also seems to us to show that policies that avoid discrimination on the grounds of race, gender, age or irrelevant disabilities can be not just socially responsible but also a shrewd recruitment policy.

A general approach might be the following:

- **Create a job specification.** Advice is needed, as there will be legal implications in an official document. However, formally or informally, the requirements of the job should be documented and agreed.
- **Create a job holder profile.** Using the job specification, a profile of the person needed to carry out the job is constructed. The qualities, qualifications, education and experience required would be listed.

- **Obtain applicants.** Typically, an advertisement would be placed, either within the organization or outside in the trade or local press. The job holder profile would be examined carefully to identify the medium most likely to reach the largest number of potential applicants at least cost. For example, if a specialist is needed it would make sense to advertise in the relevant specialist journal. The other principle is to give enough information in the advertisement to allow an element of self-elimination. By giving the salary, location, job scope and any essential qualifications, the applicants will be limited to the more realistic candidates.
- **Examine CVs.** These should be read carefully and compared to the job holder profile – nothing is more annoying for all concerned than when people have CVs which clearly indicate that they are not eligible for the job and yet they are called for interview.
- **Interviews etc.** A number of different selection techniques can be tried, including aptitude tests, personality tests, and the examination of samples of previous work. All these methods must be related to specific qualities detailed in the job holder profile. Interviews are the most commonly used method. It is better if there is more than one interview session with an applicant and with each session there should not be more than two interviewers because a greater number reduces the possibility of follow-up questions and discussion. Some formal scoring system for the qualities being judged should be devised and interviewers should then decide scores individually which are then compared. An interview should be of a technical nature where the practical expertise of the candidate is assessed, or of a more general nature if not. In the latter case, a major part of the interview will in fact be evaluating and confirming what was stated in the CV – for example any time gaps in the education and employment history would be investigated, and the precise nature of jobs previously done would need to be explored.
- **Other procedures.** References will need to be taken up where necessary, and a medical examination might be needed.

8.11 INSTRUCTION IN THE BEST METHODS

This is the second concern that we have taken from Taylor. Obviously, there is a difference between loading pig iron (one of Taylor's studies) and writing C programs but the principle of having established methods and procedures is, we hope, as well understood in software development as in steel-making.

When a new member of the team is recruited, the team leader will need to plan that person's induction into the team very carefully. Where a project is already well under way, this might not be easy. However, the effort should be made – it should pay off eventually as the new recruit will become a fully effective member of the team more quickly.

The team leader should also be aware of the need to assess continually the training needs of their team members. Just as you formulate a user requirement before considering a new system, and you construct a job holder profile before recruiting a member of staff, so a training needs profile is drawn up for each stuff member before you consider specific courses. Some training can be provided by commercial training companies. Where money is tight, other sources of training should be considered but training should not be abandoned altogether even if it consists only of a team member's being told to find out about a new software tool and then demonstrating it to colleagues. Of course the nice thing about external courses is that one gets to talk to colleagues from other organizations – but attending meetings of your local branch of one of the IS/IT professional associations can serve the same purpose.

The methods learnt need, of course, to be actually applied. Reviews and inspections should help to ensure this.

8.12 MOTIVATION

(MDU 2007-08, 2010-11, 2013-14) **(*10 Marks*)**

The third concern that we noted from Taylor was that of motivating people to work. We are now going to look at some different models of motivation that have been proposed.

The Taylorist model

Taylor's viewpoint is reflected in the use of piece-rates in manufacturing industries and sales bonuses amongst sales forces. A problem that project leaders must be aware of is that piece-rates often cause difficulties if a new system is going to change work practices. If new technology is going to improve productivity, live question of adjusting piece-rates downwards to reflect this will be a sensitive issue.

Usually, radical changes in work practices have to be preceded by a move from piece-rates to day-rates.

Even where work practices are stable and output can be easily related to reward, people paid by the amount they produce will not automatically maximize their output in order to maximize their income. The amount of output will often by constrained by 'group norms', informal, even unspoken, agreements among colleagues about the amount to be produced.

Rewards have to be related in a simple and direct way to the work produced. Where a computer system is being produced, this is not easy. It is difficult to isolate and quantify work done, especially as system development and support is very much a team effort. Typical is the sentiment expressed by one member of staff in a study of software support work practices:

'This support department does well because we're a team, not because we're all individuals. I think it's the only way the support team can work successfully.'

In this kind of environment, a reward system that makes excessive distinctions between co-workers can be damaging to morale and eventually to productivity.

Maslow's hierarchy of needs

Different people are motivated by different things. Clearly money is a very strong motivator when you are broke. However, as the basic need for cash is satisfied, other motivators are likely to emerge. Abraham Maslow, an American psychologist, suggested that there is a hierarchy of needs. As lower levels of needs are satisfied then gradually higher level needs emerge. If these are then satisfied then yet another level of need will emerge. Basic needs are for things like food and shelter. The highest level need, according to Maslow, is the need for 'self-actualization', the feeling that you are completely fulfilling your potential.

In practice, the project leader must realise that people are likely to be motivated by different things at different stages of their life. For example, salary increases, while always welcome, probably have less of an impact on the more mature employee who is already relatively well-paid, than on a new and lowly-paid trainee. Older team-members might place more value on qualities of the job such as being allowed relative autonomy when they do their work, which shows respect for their judgment and sense of responsibility.

Herzberg's two-factor theory

Certain things about a job might make you dissatisfied. If the causes of this dissatisfaction are removed, this does not necessarily make the job more exciting. On the basis of research into job satisfaction that Herzberg and his associates carried out there seemed to be two sets of factors about a job that were of importance:

- **hygiene or maintenance factors,** which can make you dissatisfied if they are not right, for example, the level of pay or the working conditions;
- **motivators,** which make you feel that the job is worthwhile, like a sense of achievement or the nature of the work itself.

The expectancy theory of motivation

A model of motivation developed by Vroom and his colleagues illustrates this. It identifies three influences on motivation:

- **expectancy,** the belief that working harder will lead to a better performance;
- **instrumentality,** the belief that better performance will be rewarded;
- **perceived value,** of the resulting reward.

Motivation will be high when all three factors are high. A zero level for any one of the factors can lead to a lack of motivation.

Imagine that you are trying to get a software package supplied by a third party to work. If you realize that you will never get it to work because of a bug in it, you will give up. No matter how hard you work you will not be able to do any better (zero expectancy).

If you are working on a package for a user and, although you think you can get it to work, you discover that the user has started employing an alternative package and no longer needs this one, then you will probably feel you are wasting your time and give up (zero instrumentality).

Given that the users really do want the package, your reward in this set of circumstances might simply be a warm feeling that you have helped your colleagues and that they are grateful to you. If in fact, when the users employ the package all they do is complain and hold you responsible for any shortcomings, then you will probably avoid getting involved if they later ask for help implementing a different package (low perceived value of reward).

The Oldham-Hackman job characteristics model

Managers should try to group together the elements of the tasks that need to be carried out so that they form meaningful and satisfying assignments. Oldham and Hackman suggest that the satisfaction that a job gives is based on five factors. The first three factors make the job 'meaningful' to the person who is doing it:

- **skill variety,** the number of different skills that the job holder has the opportunity to exercise;
- **task identity,** the degree to which your work and its results are identifiable as belonging to you;
- **task significance,** the degree to which your job has an influence on others.

  The other two factors are:
- **autonomy,** the discretion you have about the way that you do the job;
- **feedback,** the information you get back about the results of your work.

Couger and Zawacki found that programmers in general rated their jobs lower on these factors than other professions, while systems analysis and analyst-programmers rated them higher. Computer development people experienced about the same level of meaningfulness in their work as other, non-IT, professionals, but had lower perceptions of the degree of responsibility and knowledge of results of their work.

Cheney found that in the programming environment, the degree to which programmers got feedback on their work and the degree to which they could contribute to decision making had positive influences on both productivity and job satisfaction, although 'consideration', which was 'the degree to which the leader develops a work climate of psychological support, mutual trust and respect, helpfulness and friendliness', rated as less important.

In practical terms, activities should be designed so that, where possible, staff follow the progress of a particular product and feel personally associated with it.

Methods of improving motivation

- **Setting specific goals.** These goals need to be demanding and yet acceptable to staff. Involving staff in the setting of goals helps to gain acceptance for them.
- **Providing feedback.** Not only do goals have to be set but staff have to have regular feedback about how they are progressing.
- **Job design.** Jobs can be altered to make them more interesting and give staff more feeling of responsibility.

Two measures are often used to enhance job design – job enlargement and job enrichment.

- **Job enlargement.** The scope of the job is increased so that the member of staff carries out a wider range of activities. It is the opposite of increasing specialization. For example, a programmer in a maintenance group might be given responsibility for specifying minor amendments as well as carrying out the actual code changes. It is significant that Couger and Zawacki found that programmer/analysts had a higher degree of job satisfaction than programmers.
- **Job enrichment.** In this case, the job is changed so that the holder carries out tasks that are normally done at a higher, managerial, level. Staff might be given responsibility for ordering consumables, for scheduling their work or for quality control. With a programmer in a maintenance team, they might be given authority to accept requests for changes which involved less than five days' work without the need for their manager's approval.

8.13 WORKING IN GROUPS

(MDU 2013-14) (10 Marks)

Having discussed people as individuals, we move on to their place in groups. A key problem with major software projects is that they always involve working in groups, but as we have seen many people attracted to computer development find this difficult.

Formal groups can be subdivided into *command groups*, which are the departmental groupings that are seen on organization hierarchy diagrams and which reflect the formal management structure and *task groups* set up to deal with specific tasks. These call on people from different command groups and would typically be disbanded once the task has been completed.

8.14 BECOMING A TEAM

(MDU 2013-14) (10 Marks)

Simply throwing people together does not mean that they will immediately be able to work together as a team. Group feelings develop over a period of time. One suggestion is that teams go through five basic stages of development:

- **forming**—the members of the group get to know each other and try to set up some ground roles about behaviour;
- **storming**—conflicts arise as various members of the group try to exert leadership and the group's methods of operation are being established;

- **norming**—conflicts are largely settled and a feeling of group identity emerges;
- **performing**—the emphasis is now on the tasks at hand;
- **adjourning**—the group disbands.

Where people are being put together into a team for the first time, then some specific team-building exercises can be undertaken. Some organizations, for example, send their management teams off on outward bound courses.

Valuable research has gone into looking at the best mix of personalities in a project team. Belbin studied teams working together on management games using various mixes of people. He initially tried putting all the people who were most able into one group. Surprisingly, these elite teams tended to do very badly – they argued a lot and as a result important tasks were often neglected.

Belbin came to the conclusion that teams needed a balance of different types of people.

- **The chair.** Not necessarily a brilliant leader but must be good at running meetings, being calm, strong but tolerant.
- **The plant.** Someone who is essentially very good at generating ideas and potential solutions to problems.
- **The monitor-evaluator.** Good at evaluating ideas and potential solutions and helping to select the best one.
- **The shaper.** Rather a worrier, who helps to direct the team's attention to the important issues.
- **The team worker.** Skilled at creating a good working environment, for example by 'jollying people along'.
- **The resource investigator.** Adept at finding resources in terms of both physical resources and information.
- **The completer-finisher.** Good at completing tasks.
- **The company worker.** A good team player who is willing to undertake less attractive tasks if they are needed for team success.

A person can have elements of more than one type. On the other hand, about 30% of the people examined by Belbin could not be classified at all! To be a good team member you must be able to:

- time your interventions, that is, not overwhelm the others in the team;
- be flexible;
- be restrained;
- keep the common goals of the team in mind all the time.

Group performance

Are groups more effective than individuals working alone? Given the preference of many people attracted to software development for working on their own, this is an important question. In many projects, judgements need to be made about which tasks are best carried out collectively and which are best delegated to individuals to do on their own. As one manager at IBM was quoted as saying: '*Some work yields better results if carried out as a team while some things are slowed down if the work is compartmentalized on an individual basis*'. Part of the answer lies in the type of task being undertaken.

One way of categorizing group tasks is into:

- additive tasks;
- compensatory tasks;

- disjunctive tasks;
- conjunctive tasks.

Additive tasks are where the efforts of each participant are added together to get the final result, as in a gang of people clearing snow. The people involved are interchangeable.

With *compensatory tasks* the judgements of individual group members are pooled so that errors by some group members are compensated for by the inputs from others. An example of this would be where individual members of a group are asked to provide estimates of the effort needed to produce a piece of software and the results are then averaged. In these circumstances, group work is generally more effective than the efforts of individuals.

With *disjunctive tasks* there is only one correct answer. The effectiveness of the group depends on:

- someone coming up with the right answer;
- the others recognizing it as being correct.

With this type of task, the group can only be as good as its best member and no better.

Conjunctive tasks are where progress is governed by the rate of the slowest performer. Software production where different staff are responsible for different modules seems to be a prime example of this. The overall task is not completed until every participant's work is complete. In this case co-operative attitudes are productive: the team members who are more advanced need to ensure the meeting of group objectives by assisting those who are behind.

With all types of collective task, but particularly with additive ones, there is a danger of *social loafing,* where some individuals do not make their proper contribution. This can certainly occur with student group activities, but is not unknown in 'real' work environments. As one software developer has commented: '[*The contribution made to others*] *is not always recognized. Nor is the lack of any contributions... nobody points out those who fail to make any contributions. Like when there's somebody with vital skills and you ask him for help, but he doesn't provide it'.*

8.15 DECISION MAKING *(MDU 2010-11, 2012-13, 2013-14)* *(10 Marks)*

Before we can look more closely at the effectiveness with which groups can make decisions we need to look in general terms at the decision-making process.

Decisions can be categorized as being:

- **structured**, generally relatively simple, routine decisions where rules can be applied in a fairly straightforward way;
- **unstructured,** more complex and often requiring a degree of creativity.

Another way of categorizing decisions is by the amount of *risk* and *uncertainty* that is involved.

Yet another distinction is between the rational-economic model and the satisfying model. The *rational-economic* model of decision making is the basis of classical economics. It predicts, for example, that a prospective buyer of personal computer equipment will purchase goods at the lowest possible price. This assumes that the decision maker has a complete knowledge of the state of the market. In order to achieve this, days, weeks, or months could be spent phoning dealers.

Sensible people probably follow a *satisficing* approach and would look at a limited number of representative outlets to get a general idea of prices. Any potential loss of money through having missed an even lower offer would probably be offset by the savings in time, phone calls, travel and so on.

Some mental obstacles to good decision making

Many management decisions in the real world, however, made under pressure and based on incomplete information, are largely intuitive. We have to accept the role of intuition but must be aware that there are some mental obstacles to effective intuitive thinking, for example:

Faulty heuristics. Heuristics mean rules of thumb. Rules of thumb can be useful but there are dangers:

- they are based only on the information that is to hand and this can be misleading;
- they are based on stereotypes, such as accepting a Welshman into a male voice choir without an audition because of the 'well-known fact' that the Welsh are a great singing nation.

Escalation of commitment. This refers to the way that once you have made a decision it is increasingly difficult to alter it even in the face of evidence that it is wrong.

Information overload. It is actually possible to be presented with too much information so that you 'cannot see the wood for the trees'.

Group decision making

With a project team, different specialists and points of view can be brought together. Decisions made by the team as a whole are more likely to be accepted than those that are imposed upon it.

Assuming that the meetings are genuinely collectively responsible and have been properly briefed, research shows that groups are better at solving complex problems where the members of the group have complementary skills and expertise. The meeting allows them to communicate freely and to get ideas accepted.

Groups are less effective when dealing with poorly structured problems, which need creative solutions. Brainstorming techniques have been developed to help groups in this situation but research shows that people often come up with more ideas individually than in a group. Where the aim is to get the involvement of end users of a computer system, then prototyping and participatory approaches such as Joint Application Development (JAD) might be adopted.

Obstacles to good group decision making

Conflict could, in fact, be less than might be expected. Experiments have shown that people will modify their personal judgements to conform to *group norms*. These are common attitudes that are developed by a group over a period of time.

You might think that this would tend to moderate the more extreme views that some individuals in the group might hold. In fact, people in groups often make decisions that carry more risk than where they have to make the decision on their own. This is known as the *risky drift*.

Measures to reduce the disadvantages of group decision making

One method of making group decision making more efficient and effective is by training members to follow a set procedure. The *Delphi technique* endeavours to collate the judgements of a number of experts without actually bringing them face-to-face. Given a problem, the following procedure is carried out:

- the co-operation of a number of experts is enlisted;
- the problem is presented to the experts;
- the experts record their recommendations;
- these recommendations are collated and reproduced;

- the collected responses are recirculated;
- the experts comment on the ideas of others and modify their recommendations if so moved;
- if the leader detects a consensus then the process is stopped, otherwise the comments are recirculated to the experts.

The big problem with this approach used to be that because the experts could be geographically dispersed the process was time consuming.

8.16 LEADERSHIP *(MDU 2009-10, 2012-13, 2013-14) (10 Marks)*

Leadership is generally taken to mean the ability to influence others in a group to act in a particular way in order to achieve group goals. A leader is not necessarily a good manager or vice versa, because managers have other roles to play, such as those of organizing, planning and controlling.

Authorities on this subject have found it very difficult to agree a list of the common characteristics of good leaders. It would, however, seem safe to say that they seem to have a greater need for power and achievement and have more self-control and more self-confidence than others.

Leadership is based on the idea of some kind of authority or power, although leaders do not necessarily have much formal authority. This power comes from either the person's position (*position power*) or from the person's individual qualities (*personal power*) or can be a mixture of the two. Position power has been further analysed into:

- **coercive power,** the ability to force someone to do something by threatening punishment;
- **connection power,** which is based on having access to those who have power;
- **legitimate power,** which is based on a person's title conferring a special status;
- **reward power,** where the holder can confer rewards on those who carry out tasks to their satisfaction.

Personal power, on the other hand, can be further analysed into:

- **expert power,** which comes from being the person who is able to do a specialized task;
- **information power,** where the holder has access to information that others do not;
- **referent power,** which is based on the personal attractiveness of the leader.

Leadership styles

Balanced against this is the need to involve the staff in some of the decision making in order to make the best use of expertise and to gain commitment.

Attempts have been made to measure leadership styles on two axes: directive vs. permissive and autocratic vs. democratic:

- **directive autocrat** makes decisions alone with close supervision of their implementation;
- **permissive autocrat** makes decision alone but gives subordinates latitude in implementation;
- **directive democrat** makes decisions participatively but closely supervises their implementation;
- **permissive democrat** makes decisions participatively and gives subordinates latitude in implementation.

Another axis on which there have been attempts to measure management qualities has been on the degree to which a manager is *task-oriented,* that is, the extent to which the execution of the task at hand is paramount, and the degree to which the manager is concerned about the people involved (*people orientation*). It is perhaps not surprising that subordinates appear to perform best with managers who score highly in both respects.

Work environments vary according to the amount of control that can be exerted over the work. Some jobs are routine and predictable (as when dealing with batched computer output). Others may be driven by outside factors (as in the case of a help-desk) or are situations where future direction is uncertain (for example, at the early stages of a feasibility study). Where there is a high degree of uncertainty, subordinates will seek guidance from above and welcome a task-oriented management style. As uncertainty is reduced, the task-oriented manager is likely to relax and to become more people-oriented and this will have good results. People-oriented managers are better where staff can control the work they do and know what to do without referring matters to their line managers. It is then argued that if control becomes even easier the people-oriented manager will be tempted to get involved in more task-centred questions and that this may have undesirable results.

Research findings also show that where team members are relatively inexperienced a task-oriented approach is most effective. As group members mature, consideration for their personal needs and aspirations becomes more valued. Where maturity is very high, then there is no need for a strong emphasis on either of these approaches.

8.17 ORGANIZATIONAL STRUCTURES

(MDU 2013-14) (10 Marks)

Formal versus informal structures

The *formal* structure is the one that is expressed in the staff hierarchy chart. It is basically concerned with *authority,* about who has which boss. It is backed by an *informal* structure of contacts and communication that grows up spontaneously among members of staff during the course of work. When the unexpected happens it is often this system that comes into play. Over a period of time, the advantages and disadvantages of different organizational structures tend to even out – the informal organization gets built up and staff find unofficial ways of getting around the obstacles posed by the formal structure.

Hierarchical approach

The 'traditional' management structure is based on the concept of the *hierarchy* – each member of staff has only one manager, while a manager will have responsibility for several members of staff. Authority flows from the top down through the structure. A traditional concern has been with the *span of control* – the number of people that a manager can effectively control.

Staff versus line

Staff in organizations can often be divided into *line* workers who actually produce the end product and support *staff* who carry out supporting roles. In some organizations that produce software for the market or as a component of a larger product which is sold, the software specialists might be seen as part of the line. In a financial organization, on the other hand, the information systems department would probably be seen as part of the support staff.

Departmentalization

In drawing up a structure, the question of *differentiation* crops up. This is the question of how the organization is to be departmentalized. This is often based on start specialisms, product lines, categories of customer or geographical location, for example.

In the case of software development, it is usually the case that either a *functional* or a *task-oriented* approach is used. With functional departmentalization, systems analysts might be put in a group separate from the programmers. The programmers would act as a pool from which resources may be drawn for particular tasks. With a task-oriented approach, the programmers and systems analysts are grouped together in one project team. The project team might be gathered in order to implement a specific long-term project or might exist on a permanent basis to service the needs of a particular set of users.

One advantage of the functional approach is that it can lead to a more effective use of staff. Programmers can be allocated to jobs as needed and be released for other work when a particular task is completed. For instance, in a project team there are bound to be periods of greater and lesser coding activity and programmers might find there are spells when they are under-utilized. The functional organization will also make it easier for programmers to have careers that are technically oriented – there will probably be a career structure within the software development department that allows the programmer to rise without having to change specialism. This type of organization should also encourage the interchange of new technical ideas among technical staff and the promulgation of company wide standards.

A disadvantage is that having two separate departments can lead to communication problems, especially if a programmer is unfamiliar with the application area. There will also be problems with software maintenance – here it is helpful to have programmers who have built up a familiarity with particular parts of the application software. Users might prefer the established project team approach because, when they require new software features, they will already have a group dedicated to their needs and will not find themselves in the position of always having to fight other departments for development resources. The project team structure tends to favour a pattern of career progression where programmers eventually become systems analysts.

A third method of departmentalization is based on lifecycle phase. Here there are separate teams for development and maintenance. Some staff can concentrate in a focused and sustained manner on developing new applications with few interruptions, while other teams, more oriented towards service and support, deal with maintenance.

Some organizations have attempted to get the best of all worlds by having a *matrix* structure. In this case the programmer would have two managers: a project leader who would give day-to-day direction about the work in hand and a programming manager who would be concerned about such things as career development.

Centralized versus decentralized group structures

At the level of a project group, a decentralized organization would mean that the group members would tend to make major decisions collectively and that there would be a large degree of free communication among group members. With the centralized approach the group would be broken down into sections, each of which would be directed by a leader who communicates on behalf of the section with other groups.

Decentralized groups, because of the time taken to debate things, tend to work more slowly. They are likely to be affected by the establishment of group norms and the influence of the risky shift, which has already been described. However, they are better at dealing with complex problems while the centralized group organization deals more effectively with simple problems.

The discussion of centralized versus decentralized groups assumes that software development work has to be done as a group. In fact, given the preference of many software developers for working on their own, an organization where each programmer works in isolation can be envisaged – indeed there are software houses that are based on people working at home.

Egoless programming

In the early days of computer development, managers tended to think of the programmer as communing mysteriously with the machine. The tendency was for programmers to see programs as being an extension of themselves and to feel over-protective towards them. The effects of this on the maintainability of programs can be imagined. Gerald Weinberg made the then revolutionary suggestion that programmers and programming team leaders should read other people's programs. Programs would become in effect the common property of the programming group and programming would become 'egoless'. Peer code reviews are based on this idea. Weinberg's ideal programming team was a decentralized group freely communicating within itself.

Chief programmer teams

The larger the decentralized group, the slower it will get, because of the increased communication. On really large time-critical projects, a more formalized centralized structure is essential. Brooks pointed out the need for design consistency when producing a large complex system and how this might be difficult when there are a large number of people involved in producing a piece of software. One suggestion was to try to reduce the number of people actually creating software but to make these programmers as productive as possible by giving them as much support as possible.

The result of this train of thought was the *chief programmer* team. The chief programmer is the person who defines the specification, and designs, codes, tests and documents the software. There is also a *copilot,* with whom the chief programmer can discuss problems and who writes some code. They are supported by an *editor* to write up the documentation drafted by the chief programmer, a *program clerk* to maintain the actual code, and a *tester.* The general idea is that this team is under the control of a single unifying intellect.

The chief programmer concept was used on the influential *New York Times* data bank project, where many aspects of structured programming were tried out. In this case, each chief programmer managed a senior level programmer and a program librarian. Additional members could be added to the team on a temporary basis to deal with particular problems or tasks.

The problem with this kind of organization is getting hold of really outstanding programmers to carry out the chief programmer role. There are also the dangers of information overload on the chief programmer, and of staff dissatisfaction among those who are there simply to minister to the needs of the superstar chief programmers.

Controlled decentralized groups

This compromise structure has been suggested and seems to follow common industry practice. A project team is made of groups under the leadership of senior programmers. Within these groups there is free communication and a practice of reviewing each others' work. Communication with other groups is at senior programmer level, while a project leader has overall authority.

CONCEPTUAL SHORT QUESTIONS WITH ANSWERS

1. What is the role of the document editor?

Ans. The document editor is responsible for the low-level integration of the document. Also, he is responsible for the overall format of the document and its index.

2. What do you understand by 'contracting'?

Ans. Contracting means Outsourcing.

3. When the effort of each participant is being added, the task is called as?

Ans. Additive Task.

4. What is the purpose of Configuration manager?

Ans. The configuration manager is responsible for maintaining a revision history of the document. Also, he is responsible for maintaining traceability information.

5. When more people are added to a project, what happens to the communication paths?

Ans. As more people are added to work on a project, the number of different communication paths increases rapidly.

It increases according to the formula: $\frac{n(n-1)}{2}$ where 'n' is the number of people on the project team.

REVIEW QUESTIONS

1. When more people are added to a project, what happens to the communication paths.
2. What is the purpose of configuration manager?
3. List and discuss the responsibilities of the software managers.
4. (*a*) Discuss different types of contracts, its stages and typical terms in it. Also describe the term contract management.

   (*b*) Consider a situation where you are a manager of a software company. How will you prove your leadership in the company? Discuss in detail.
5. Define contract and contract management.
6. For successful completion of the project, it is necessary to choose right kind of people, retain them, to instinct them properly and to maintain cohesiveness within the teams. How all this can be achieved?
7. Define human resource management and put a light on the following aspects:

   (*a*) Creating team spirit

   (*b*) Motivating people.
8. Effective SPM practices focus on four major factors:

   People, Product, Process and Project

   Explain the ways in which each of these be handled so that the project is accomplished in proper manner.

CHAPTER 9

SOFTWARE QUALITY

INSIDE THIS CHAPTER

9.1 INTRODUCTION

Software quality and software quality assurance (SQA) are the most widely talked issues in the software industry today. The aim of SQA is to assist software development organisations develop high quality software products in a repeatable way so as to make the software development process person independent (Ince 1991; Kitchenham 1996; Ince 1994). Total quality management for software is gaining importance (Sckulmeyer 1996).

We examine the qualities that are pertinent to software products and software production processes. These qualities will become our goals in the practice of software engineering.

While quality is generally agreed to be 'a good thing', in practice the quality of a system will be a vague, undefined, attribute if we are not careful. We therefore need to define precisely what qualities we require of a system. However, this by itself is not enough—we need to be able to judge objectively whether a system meets our quality requirements and this leads to the need for measurement.

The basic goal of software engineering is to produce quality software. Software quality is a broad and important field of software engineering addressed by several standardization bodies such as ISO, IEEE, ANSI etc.

This leads to a slightly different emphasis rather than concentrating on the quality of the final system, a potential customer for software might try to check that the suppliers were using the best methods.

This chapter examines these issues.

9.2 THE PLACE OF SOFTWARE QUALITY IN PROJECT PLANNING

(MDU 2008-09, 2011-12) (10 Marks)

Quality will be of concern at all stages of project planning and execution, but will be of particular interest at the following points in the Step Wise framework (Fig. 9.1).

- **Step 2: Identify project infrastructure.** Within this step, activity 2.2 identifies installation standards and procedures. Some of these will almost certainly be about quality.

- **Step 3: Analyse project characteristics.** In activity 3.2 ('Analyse other project characteristics—including quality based ones') the system to be implemented will be examined to see if it has any special quality requirements.
- If, for example, it is extremely safety-critical then a whole range of additional activities can be added; these include such things as *n*-version development, where a number of teams develop versions of the same software that are then run in parallel with the outputs being cross-checked for discrepancies.
- **Step 4: Identify the products and activities of the project.** It is at this point that the entry, exit and process requirements are identified for each activity. The nature of these requirements is described later in this chapter.

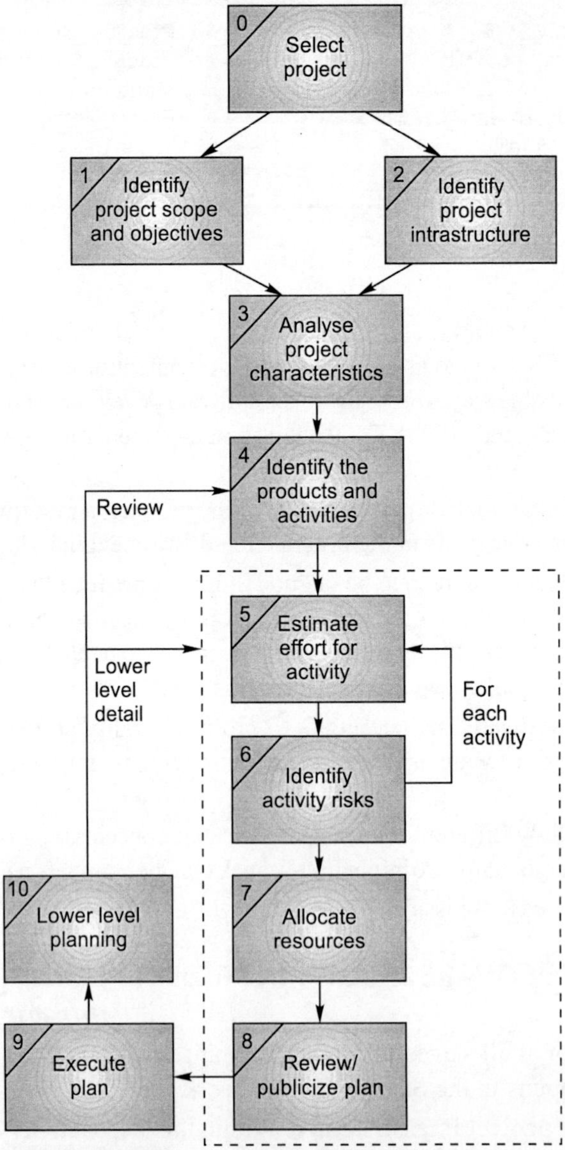

Fig. 9.1 The place of software quality in Step Wise.

- **Step 8: Review and publicize plan.** At this stage, the overall quality aspects of the project plan are reviewed.

9.3 THE IMPORTANCE OF SOFTWARE QUALITY *(MDU 2012-13) (10 Marks)*

We would expect quality to be a concern of all producers of goods and services. However, the special characteristics of software, and in particular its intangibility and complexity, make special demands.

- **Increasing criticality of software.** The final customer or user is naturally anxious about the general quality of software, especially its reliability. This is increasingly the case as organizations become more dependent on their computer systems and software is used more and more in areas that are safety-critical, for example, to control aircraft.
- **The intangibility of software.** This makes it difficult to know that a particular task in a project has been completed satisfactorily. The results of these tasks can be made tangible by demanding that the developer produce 'deliverables' that can be examined for quality.
- **Accumulating errors during software development.** As computer system development is made up of a number of steps where the output from one step is the input to the next, the errors in the earlier deliverables will be added to those in the later steps leading to an accumulating detrimental effect. In general, the later in a project that an error is found the more expensive it will be to fix. In addition, because the number of errors in the system is unknown, the debugging phases of a project are particularly difficult to control.

For these reasons quality management is an essential part of effective overall project management.

9.4 DEFINING SOFTWARE QUALITY *(MDU 2010-11, 2012-13, 2013-14) (10 Marks)*

Software quality is a board and important field of software engineering addressed by several standardization bodies such as ISO (especially, in ISO 9000), ANSI, IEEE, etc. (Smith 1990).

The definition of the ISO 8204 for quality:

"*Totality of characteristics of an entity that bears on its ability to satisfy stated and implied needs*"

This means that the software product delivered should be as per the requirements defined.

OR

User satisfaction = Complaint product + Good quality + Delivery within budget and schedule.

—Robert Glass

Quality is a rather vague term and we need to define carefully what we mean by it. For any software system, there should be three specifications:

- a functional specification describing what the system is to do;
- a quality (or attribute) specification concerned with how well the functions are to operate;
- a resource specification concerned with how much is to be spent on the system.

Methodologies such as SSADM produce primarily functional requirements—so how are the required qualities of the proposed system, such as its flexibility when change is needed, to be defined?

There have been several attempts to identify specific product qualities that are appropriate to software. James A. McCall, for instance, grouped software qualities into three sets:

- product operation qualities;
- product revision qualities;
- product transition qualities.

The definitions below are those given by McCall, but the reader will come across others. In particular circumstances, additional qualities are of interest.

Product operation quality factors

- **Correctness.** The extent to which a program satisfies its specifications and fulfils the user's objectives.
- **Reliability.** The extent to which a program can be expected to perform its intended function with required precision.
- **Efficiency.** The amounts of computer resources required by the software.
- **Integrity.** The extent to which access to software or data by unauthorized persons can be controlled.
- **Usability.** The effort required to learn, operate, prepare input and interpret output.

Product revision quality factors

- **Maintainability.** The effort required to locate and fix an error in an operational program.
- **Testability.** The effort required to test a program to ensure it performs its intended function.
- **Flexibility.** The effort required to modify an operational program.

Product transition quality factors

- **Portability.** The effort required to transfer a program from one hardware configuration and/ or software system environment to another.
- **Reusability.** The extent to which a program can be used in other applications.
- **Interoperability.** The effort required to couple one system to another.

McCall's software quality factors reflect the external view of software that users would have. For instance, usability would be a key concern of users. These quality factors have to be translated into internal factors of which the developers would be aware—*software quality criteria* (Table 9.1).

Table 9.1: Software quality criteria

Quality factor	*Software quality criteria*
Correctness	traceability, consistency, completeness
Reliability	error tolerance, consistency, accuracy, simplicity
Efficiency	execution efficiency, storage efficiency
Integrity	access control, access audit
Usability	operability, training, communicativeness, input/output volume, input/output rate
Maintainability	consistency, simplicity, conciseness, modularity, self-descriptiveness
Testability	simplicity, modularity, instrumentation, self-descriptiveness
Flexibility	modularity, generality, expendability, self-descriptiveness
Portability	modularity, self-descriptiveness, machine independence, software system independence
Reusability	generality, modularity, software system independence, machine independence, self-descriptiveness
Interoperability	modularity, communications commonality, data commonality

Defining quality is not enough. If we are to judge whether a system meets our requirements we need to be able to measure its qualities. For each criterion, one or more measures have to be invented to assess the degree to which the quality is present.

Any good relative measure must be able to relate the number of units to the maximum possible in the circumstances. The maximum number of faults in a program, for example, is going to be related to the size of the program, so a measure of 'faults per thousand lines of code' is more helpful than 'total faults in a program' as a means of judging the quality of a program.

Trying to find measures for a particular quality helps to clarify ideas about what that quality really is. What is being asked is, in effect, 'how do we know when we have been successful?' An answer to this is essential if the quality objectives are to be communicated to a large number of people.

In some cases we can measure the quality directly while in other cases the thing being measured is not the quality itself but an indicator of the degree to which the quality is present. By identifying measures, management are setting targets for project team members so care has to be taken that an improvement in the measured attribute is always going to be valid. For example, the number of errors found in program inspections could be counted. This count could, of course, be improved by allowing more errors to go through to the inspection stage rather than eradicating them earlier—which is not quite the point!

In general, the user of software would be concerned with measuring what MeCall called *quality factors* while the developers would be concerned with *quality criteria.*

The following should be laid down for each quality:

- **scale:** the unit of measurement;
- **test:** the practical test of the extent to which the attribute quality exists;
- **worst:** the worst acceptable value;
- **plan:** the value that it is planned to achieve;
- **best:** the best value that appears to be feasible (the 'state of the art' limit); this would be a level that is known to have been achieved elsewhere;
- **now:** the value that applies currently.

In order to derive these *quality specifications* it is often necessary to break down a quality criterion into further sub-criteria. Take the quality criterion 'communicativeness', which contributes to the quality factor 'usability'. One aspect of this might be the ease of understanding of the menu structure, in particular, how easy it is to find the command to carry out some function. Another aspect of communicativeness would be how informative the error messages were, while yet another would be the clarity of the 'help' pages.

9.5 ISO 9126 *(MDU 2008-09, 2011-12, 2012-13, 2013-14)* *(10 Marks)*

Over the years, various lists of software quality characteristics have been put forward, such as those of McCall, described above, and of Boehm. A difficulty has been the lack of agreed definitions of the qualities of good software. The term 'maintainability' has been used, for example, to refer to the ease with which an error can be located and corrected in a piece of software, and also in a wider sense to include the ease of making any changes. For some, 'robustness' has meant the software's tolerance of incorrect input, while for others it has meant the ability to change program code without introducing unexpected errors. ISO 9126 standard was published in 1991 to tackle the question of the definition of software quality. This 13 page document was designed as a foundation upon which further, more detailed, standards could be built.

ISO 9126 identifies six software quality characteristics:

- **Functionality.** Which covers the functions that a software product provides to satisfy user needs;
- **Reliability.** Which relates to the capability of the software to maintain its level of performance;
- **Usability.** Which relates to the effort needed to use the software;
- **Efficiency.** Which relates to the physical resources used when the software is executed;
- **Maintainability.** Which relates to the effort needed to the make changes to the software;
- **Portability.** Which relates to the ability of the software to be transferred to a different environment.

ISO 9126 suggests sub-characteristics for each of the primary characteristics. It is perhaps indicative of the difficulties of gaining widespread agreement that these sub-characteristics are outside the main standard and are given in the document for information only. They are useful as they clarify what is meant by the main characteristics.

Characteristic	*Sub-characteristics*
Functionality	Suitability
	Accuracy
	Interoperability
	Compliance
	Security

Compliance refers to the degree to which the software adheres to application-related standards or legal requirements. Typically, these could be auditing requirements.

Interoperability and *security* are good illustrations of the efforts of ISO 9126 to clarify terminology. Interoperability refers to the ability of the software to interact with other systems. The framers of ISO 9126 have chosen this word rather than *compatibility* because the latter causes confusion with the characteristic referred to by ISO 9126 as *replaceability* (see below).

Characteristic	*Sub-characteristics*
Reliability	Maturity
	Fault tolerance
	Recoverability

Maturity refers to the frequency of failure due to faults in a software product, the implication being that the more the software has been used, the more faults will have been uncovered and removed. It is also interesting to note that *recoverability* has been clearly distinguished from *security,* which describes the control of access to a system.

Characteristic	*Sub-characteristics*
Usability	Understandability
	Learnability
	Operability

Understandability is a clear quality to grasp, although the definition 'attributes that bear on the users' efforts for recognizing the logical concept and its applicability' in our view actually makes it less clear!

Note how *learnability* has been distinguished from *operability*. A software tool might be easy to learn but time-consuming to use because, say, it uses a large number of nested menus. This is fine for a package that is used only intermittently, but not where the system is used for several hours each day by the end user. In this case, learnability has been incorporated at the expense of operability.

Characteristic	*Sub-characteristics*
Efficiency	Time behaviour
	Resource behaviour
Maintainability	Analysability
	Changeability
	Stability
	Testability

Analysability is the quality that McCall called *diagnosability*, the ease with which the cause of a failure can be determined. *Changeability* is the quality that others have called *flexibility*: the latter name is perhaps a better one as changeability has a slightly different connotation in plain English it implies that the suppliers of the software are always changing it!

Stability, on the other hand, does not mean that the software never changes: it means that there is a low risk of a modification to the software having unexpected effects.

Characteristic	*Sub-characteristics*
Portability	Adaptability
	Installability
	Conformance
	Replaceability

Conformance, as distinguished from *compliance*, relates to those standards that have a bearing on portability. The use of a standard programming language common to many software/hardware environments is an example of conformance. *Replaceability* refers to the factors that give 'upwards compatibility' between old software components and the new ones, 'downwards compatibility' is specifically excluded from the definition.

ISO 9126 provides some guidelines for the use of the quality characteristics. The fact that the relative importance of different quality characteristics will depend on the type of product under examination is stressed. Thus reliability will be of particular concern with safety-critical systems while efficiency will be crucial for some real-time systems. For interactive end user systems, the key quality might be usability. Once the requirements for the software product have been established, the following steps are laid down:

Quality metrics selection. Measurements that correlate to the characteristics of each quality have to be identified. No specific guidance is given by the ISO 9126 standard on the applicability of the various measurements that might be used.

Ratings level definition. The metrics used must be mapped onto scales that indicate the degree to which the requirements have been satisfied. For example, in one application 'time behaviour' in the sense of response time might be important. For a key transaction, actual response times might be mapped onto quality scores:

response time (seconds)	*quality score*
< 2	5
2-3	4
4-5	3
6-7	2
8-9	1
> 9	0

Assessment criteria definition. The way that the quality scores are combined or summarized to give an overall view of the product has to be defined. The software product has now to be evaluated by measuring its qualities, converting them to quality scores or ratings, and summarising the ratings to obtain an overall judgement. ISO 9126 does not specify how this has to be done, only that some method must be devised.

One approach, for example, recognizes that some quality rating levels will be mandatory. If a product fails to reach any of the mandatory rating levels it must be rejected regardless of how good it might be in other ways. Other characteristics can be desirable but not essential. For these desirable characteristics it is possible to give a rating in the range 1–5, say, reflecting how important they are. Above we have shown in the case of time behaviour how its quality in a particular product can be awarded a quality score on a scale 0 to 5. The scores for the more or less important qualities can be given due weight by multiplying each one by its importance weighting. These weighted scores can then be summed to obtain an overall score for the product. The scores for various products can then be compared to get a first cut order of preference, for example, the quality of two products might need to be compared on the grounds of usability, efficiency and maintainability. The importance of each of these qualities might be rated as 3, 4 and 2 respectively, out of a possible maximum of 5. Quality tests could result in the scores shown in Table 9. 2.

Table 9.2: Quality rating scores

Product quality	*Importance rating (a)*	*Product A*		*Product B*	
		quality score (b)	*weighted score (a × b)*	*quality score (c)*	*weighted score (a × c)*
usability	3	1	3	3	9
efficiency	4	2	8	2	8
maintainability	2	3	6	1	2
overall			17		19

An attempt was made to use the ISO 9126 standard to define the qualities needed in the software to control systems in an unmanned station in the Antarctic that was to perform scientific experiments that were to be monitored and controlled by satellite link from Italy. Useful advice that came out of this was that the qualities required can vary from component to component within the system and that therefore quality definition at the level of the overall system is not always appropriate. The researchers also found that the most important quality for this application was availability, which was a mixture of the ISO 9126 top level qualities of reliability and maintainability separating the two elements was in practice very difficult.

9.6 PRACTICAL SOFTWARE QUALITY MEASURES (Important)

Below are some ways of measuring particular qualities. It is emphasized that the measures are illustrations only and should certainly not be treated as definitive! Each project will need to have its own measures devised to meet its own specific needs. The measures described relate to the final software products of a project.

Reliability

This might be measured in terms of:

- **availability,** the percentage of a particular time interval that a system is usable;
- **mean time between failures,** the total service time divided by the number of failures;
- **failure on demand,** the probability that a system will not be available at the time required or the probability that a transaction will fail;
- **support activity,** the number of fault reports that are dealt with.

Maintainability

This is closely related to flexibility, the ease with which the software can be modified. The main difference is that before an amendment can be made, the fault has to be diagnosed. Maintainability can therefore be seen as flexibility plus a new quality, diagnosability, which might be defined as the average amount of time needed to diagnose a fault.

Extendibility

This is a component of the more general quality of flexibility. It can be defined as the productivity needed to incorporate a new feature into an existing system expressed as a percentage of the normal productivity when developing the software from scratch.

9.7 PRODUCT VERSUS PROCESS QUALITY MANAGEMENT

(MDU 2009-10, 2012-13, 2013-14) (10 Marks)

The measurements described above can be taken only after the system is operational. It might then be too late to do anything to remedy problems.

The system development process is made up of a number of activities that are linked together so that the output from one activity is the input to the next (Fig. 9.2). Thus, program testing will depend on there being a program to test that will be the product of the program coding stage. Errors can enter the process at any stage. They can be introduced either because of a defect in the way a process is carried out, as when programmers make mistakes in the logic of their programs, or because information has not been passed clearly and unambiguously between stages.

Errors that creep in at the early stages are more expensive to correct at later stages, for the following reasons.

- The later the error is found the more rework at more stages of development will be needed. If an error in the specification is found at the testing stage, then this will mean rework at all the stages between specification and testing.
- The general tendency is for each successive stage of development to be more detailed and less able to absorb change.

Errors should therefore be eradicated by careful examination of the products of each stage before they are passed on to the next. To do this, the following *pax process requirements* should be specified for each activity.

- **Entry requirements,** which have to be in place before an activity can start. An example would be that a comprehensive set of test data and expected results be prepared and approved before program testing can commence.
- **Implementation requirements,** which define how the process is to be conducted. In the testing phase, for example, it might be laid down that whenever an error is found and corrected, all test runs must be repeated, even those that have previously been found to run correctly.
- **Exit requirements,** which have to be fulfilled before an activity is deemed to have been completed, for example, for the testing phase to be recognized as being completed, all tests will have to have been run successfully with no outstanding errors.

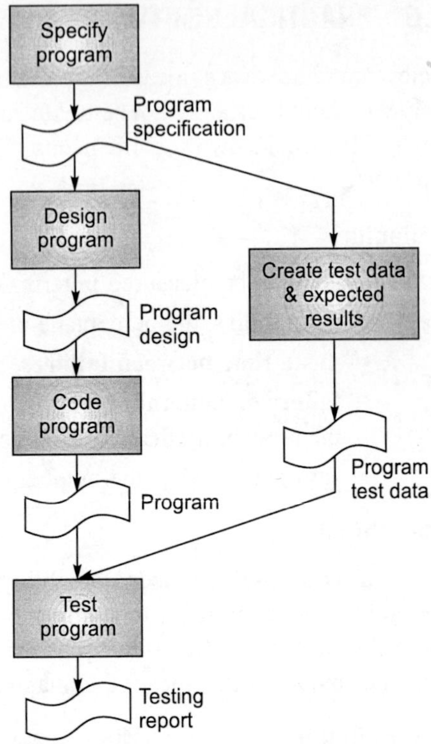

Fig. 9.2 An example of the sequence of processes and deliverables.

9.8 EXTERNAL STANDARDS

BS EN ISO 9001

It is now common to include in contracts terms covering the types off technique that a contractor will use. Various national and international standards bodies, including the British Standards Institution (BSI in the United Kingdom, have inevitably become involved in the creation of standards for quality management systems. The British standard is now called BS EN ISO 9001:1994, which is identical to the international standard, ISO 9001:1994. Standards such as the ISO 9000 series aim to ensure that a monitoring and control system to check quality is in place. They are concerned with the certification of the development process, not of the end product, as in the case of crash helmets and electrical appliances with their familiar CE labels. The ISO 9000 series govern quality systems in general terms and not just those in the software development environment.

There has been some controversy over the value of these standards. Stephen Halliday, writing in *The Observer,* had misgivings that these standards are taken by many customers to imply that the final product is of a certified standard although as Halliday says, 'It has nothing to do with the quality of the product going out of the gate. You set down your own specifications and just have to maintain them, however low they may be'. It has also been suggested that obtaining certification can be an expensive and time-consuming process that can put smaller, but still well-run, businesses at a disadvantage. Finally, there has been a concern that a preoccupation with certification might distract attention from the real problems of producing quality products.

Putting aside these reservations, let us examine how the standard works. A primary task is to identify those things that are to be the subject of quality requirements. Having defined the requirements, a system must be put in place to check that the requirements are being fulfilled and that corrective action is being taken where necessary.

An overview of BS EN ISO 9001 QMS requirements

In order for a quality management system (QMS) to meet the standard it has to conform to certain requirements which are summarized below:

(*a*) The management must define and document the policy concerning quality and must ensure that this policy is communicated to all levels of the organization.

(*b*) All quality control procedures must be documented.

(*c*) All contracts to supply goods or services must contain mutually agreed requirements that the developer is capable of delivering.

(*d*) There must be procedures to control and verify the design of the system to be supplied so that it meets the requirements agreed with the customer.

(*e*) There must be procedures to approve design and other documentation.

(*f*) Where components of the system to be supplied to the client are obtained from third parties there must be procedures to ensure, check and maintain the quality of these components.

(*g*) Individual products must be identifiable as should their components.

(*h*) The process by which the final product is created must be planned and monitored.

(*i*) Inspection and testing must take place during the development phase, at its completion and before delivery. Tests and inspections must also be carried out on components obtained from third parties.

(*j*) The equipment used in the production process itself must be properly controlled with respect to quality.

(*k*) The testing status of all components and systems must be clearly recorded at all times.

(*l*) Care must be taken to ensure that items that are known to be defective are not inadvertently used.

(*m*) When a defect is detected, measures must be undertaken to remove the defective part and to ensure that the defect does not occur again.

(*n*) Satisfactory procedures must be in place to deal with correct handling, storage, packaging and delivery of the product.

(*o*) Sufficient records must be maintained to demonstrate that the quality system is working satisfactorily.

(*p*) The software quality management system must be audited on a regular basis.

(*q*) Servicing and support activities must be subject to the quality management system.

(*r*) The developer must establish appropriate statistical techniques to verify the acceptability of the final product.

Tick IT

The ISO 9000 standards refer to quality management systems in general but in the United Kingdom, the Department of Trade and Industry (DTI) have formulated the TickIT standards which give an interpretation of these standards, which applies specifically to software development. This includes such requirements as:

- a detailed development plan is required before development is embarked upon;
- change control procedures should be used at all stages of development;
- design reviews must take place;
- the suitability of the design methodology must be reviewed;
- progress must be reviewed on a systematic basis;
- it must be possible to trace back the features of software design to specifications and requirements;
- designs must be properly documented;
- suitable test plans, specifications and records must be produced;
- a code of practice must be in place which governs the way the software is developed.

The code of practice must include the requirements that:

- the design must be broken down into levels, each with identifiable inputs and outputs;
- software must be organized into modules;
- a module must normally perform a single function or a set of related functions;
- a plain language description must exist for each module.

A Tick IT auditor can certify that a particular organization conforms to these standards. This is called *certification.* The bodies doing this certification have to be *accredited* by the National Council for Certification Bodies (NACCB) on behalf of the DTI. The scheme is now administered by DISC, a part of the British Standards Institution.

Capability process models

Rather than just checking that a system is in place to detect faults, a customer might wish to check that a supplier is using software development methods and tools that are likely to produce good quality software. Even the TickIT recommendations can be regarded as fairly minimal. A customer will feel more confident, for instance, if they know that the software supplier is using structured methods. In the United States, an influential *capability maturity model* (CMM) has been developed at the Software Engineering Institute (SEI), a part of the Carnegie-Mellon University. This attempts to place organizations producing software at one of five levels of process maturity to indicate the sophistication and quality of their software production practices. These levels are defined as follows:

- **Level 1: Initial.** The procedures followed tend to be haphazard. Some projects will be successful, but this tends to be because of the skills of particular individuals including project managers. The is no level 0 and so any organization would be at this level by default.
- **Level 2: Repeatable.** Organizations at this level will have basic project management procedures in place. However, the way an individual task is carried out will depend largely on the person doing it.
- **Level 3: Defined.** The organization has defined the way in which each task in the software development life cycle is to be done.
- **Level 4: Managed.** The products and processes involved in software development are subject to measurement and control.
- **Level 5: Optimizing.** Improvement in procedures are designed and implemented using the data gathered from the measurement process.

For each of the levels, apart from the default level 1, *key process areas* (KPAs) have been identified as distinguishing the current level from the lower ones. These are listed in the Table 9.3.

The assessment is done by a team of assessors coming into the organization and interviewing key staff about their practices using a standard questionnaire to capture the information. A key objective is not just to assess, but to recommend specific actions to bring the organization up to a higher level.

A criticism has been made of the approach that it is unrealistic to try and assess an organization as a whole—in reality there will be major differences between the way that individual projects are conducted. *Bootstrap,* which is a European initiative along the same lines as CMM, does allow assessment to be done at a project level.

Table 9.3: CMM Key process areas

Level	*Key process areas*
1. Initial	not applicable
2. Managed	configuration management, quality assurance, sub-contract management, project tracking and oversight, project planning
3. Defined	peer reviews, inter-group coordination, software product engineering, integrated software management, training programme, organization process definition and focus
4. Managed	quality management, process measurement and analysis
5. Optimizing	process change management, technology innovation, defect prevention

Assessing software products

The concern in this section has so far been with the assessment of organizations and the processes that they use to produce software, but many purchasers of software, including project managers contemplating the purchase of software tools are more directly worried about the quality of the software product itself. Compilers for some programming languages for example, are subject to certification. Much progress, however, has still to be made in this area.

9.9 TECHNIQUES TO HELP ENHANCE SOFTWARE QUALITY *(MDU 2013-14) (10 Marks)*

So far in this chapter we have looked at the steps a customer might take to ensure the quality of software produced by an outside supplier. We now need to look at what techniques a project team might wish to employ to help them improve their own software development processes. Three main themes emerge:

- **Increasing visibility.** A landmark in this movement towards making the software development process more visible was the advocacy by the American software guru, Gerald Weinberg, of 'egoless programming'. Weinberg encouraged the simple practice of programmers looking at each other's code.
- **Procedural structure.** At first programmers were more or less left to get on with writing the programs, although there might be some general guidelines. Over the years there has been the growth of methodologies where every process in the software development cycle has carefully laid down steps.
- **Checking intermediate stages.** It seems inherent in human nature to push forward quickly with the development of any engineered object until a 'working' model, however imperfect, has been produced that can then be 'debugged'. One of the elements of the move towards quality practices has been to put emphasis on checking the correctness of work at its earlier, conceptual, stages.

We are now going to look at some specific techniques in more detail. The push towards more visibility has been dominated by the increasing use of walkthroughs, inspections and reviews. The movement towards more procedural structure inevitably leads to discussion of structured programming techniques and to its later manifestation in the ideas of 'clean-room' software development.

The interest in the dramatic improvements made by the Japanese in product quality has led to much discussion of the quality techniques they have adopted such as the use of quality circles and these will be looked at briefly. Some of these ideas are variations on the theme of inspection and clean-room development, but they are seen from a slightly different angle.

Inspections

The principle of inspection can be extended to any document that is produced at any stage of the development process, for instance, lest data needs to be reviewed its production is usually not a high profile task even though errors can get through to operational running because of its poor quality.

The underlying procedure with inspections is that when a piece of work is completed, copies of the work are distributed to co-workers who then spend sometime going through the work, noting any defects. A meeting is then held where the work is discussed and a list of defects requiring rework is produced, the work to be examined could be, for instance, a program listing that is free of compilation errors.

Our own experience of using this technique has been that:

- it is a very effective way of removing superficial errors from a piece of work;
- it motivates the programmer to produce a better structured and self-explanatory program because they know that other people will be criticizing it;
- it helps spread good programming practices because the participants discuss the advantages and disadvantages of specific pieces of code;
- it can enhance team spirit.

The item will usually be reviewed by colleagues who are involved in the same area of work, so that a programmer's work, for example, will be reviewed by fellow programmers. However, to reduce the problems of incorrect communication between different stages there will be representatives from the stages that precede and follow the one that produced the work under review.

IBM have put the review process on a much more structured and formal basis, and have produced statistics to show its effectiveness. A Fagan inspection (named after the IBM employee who pioneered the technique) is a much more formalized procedure which is led, not by the author of the work, but by a specially trained 'moderator'.

The general principles behind the Fagan method

- Inspections are carried out on all major deliverables.
- All types of defect are noted not just logic or function errors.
- Inspections can be carried out by colleagues at all levels except the very top.
- Inspections are carried out using a predefined set of steps.
- Inspection meetings do not last for more than two hours.
- The inspection is lead by a moderator who has had specific training in the technique.
- The other participants have defined roles. For example, one person will act as a recorder and note all defects found and another will ad as reader and take the other participants through the document under inspection.

- Checklists are used to assist the fault-finding process.
- Material is inspected at an optimal rate of about 100 lines an hour.
- Statistics are maintained so that the effectiveness of the inspection process can be monitored.

Structured programming and clean-room software development

One of the people most closely associated with the origins of structured programming is Dijkstra. In the late 1960s, software was seen to be getting more complex while the capacity of the human mind to hold detail remained limited. It was also realized that it was impossible to test any substantial piece of software completely—there were just too many possible combinations of inputs. The most that testing could do was prove the presence of effort, not their absence. It was suggested by Dijkstra and others that, because of this, the only way that we could reassure ourselves about the correctness of software was by actually looking at the code.

The way to deal with complex systems, it was contended, was to break them down into components that were of a size for the human mind to comprehend. For a large system there would be a hierarchy of components and sub-components. For this decomposition to work properly, each component would have to be self-contained with only one entry and exit point.

The ideas of structured programming have been further developed into the ideas of clean-room software development by people such as Harlan Mills of IBM. With this type of development there are three separate teams:

- a specification team, which obtains the user requirements and also a usage profile estimating the volume of use for each feature in the system;
- a development team, which develops the code but does no machine testing of the program code produced;
- a certification team, which carries out testing.

Any system is produced in increments, each of which should be capable of actual operation by the end user. The development team does no debugging; instead, all software has to be verified by them using mathematical techniques The argument is that software that is constructed by throwing up a crude program that then has test data thrown at it and a series of hit-and-miss amendments made to it until it works is bound to be unreliable.

The certification team carry out the testing, which is continued until a statistical model shows that the failure intensity has been reduced to an acceptable level.

Formal methods

In the section above on clean-room development, the use of mathematical verification techniques was mentioned. These techniques use unambiguous, mathematically-based, specification languages of which Z and VDM are examples. They are used to define *pre-conditions* and *post-conditions* for each procedure. Pre-conditions define the allowable states, before processing, of the various items of data that a procedure is to work upon. The post-conditions define the state of those data items after the procedure has been executed. Because the mathematical notation is precise, a specification expressed in this way should be unambiguous. It should also be possible to prove mathematically (in much the same way that at school you learnt to prove Pythagoras' theorem) that a particular algorithm will work on the data defined by the pre-conditions in such a way as to produce the post-conditions. It needs hardly be said that in many cases this will be more easily said than done. In fact structured programming can be seen as an attempt to analyse a program structure into small, self-contained, procedures that will be amenable to this formal verification approach.

Software quality circles

Much interest has been shown in Japanese software quality practices. The aim of the 'Japanese' approach is to examine and modify the activities in the development process in order to reduce the number of errors that they have in their end products. Testing and Fagan inspections can assist the removal of errors but the same types of error generally occur again and again in successive products created by a specific type of process. By uncovering the source of errors, this repetition can be eliminated. To do this needs the involvement of all the staff in identifying the causes of errors.

Staff are involved in the identification of sources of errors through the formation of *quality circles*. These can be set up in all departments of an organization including those producing software where they are known as *software quality circles* (SWQC).

A quality circle is a group of four to ten volunteers working in the same area who meet for, say, an hour a week to identify, analyse and solve their work-related problems. One of their number is the group leader and there is likely to be an outsider, *a facilitator,* who can advise on procedural matters. In order to make the quality circle work effectively training needs to be given.

Problem solving by quality circles

The steps that the circle go through in solving problems are:

(*a*) identify a list of problems;
(*b*) select one problem to solve;
(*c*) clarify the problem;
(*d*) identify and evaluate the causes;
(*e*) identify and evaluate the solutions;
(*f*) decide on a solution;
(*g*) develop an implementation plan;
(*h*) present the plan to management;
(*i*) implement the plan;
(*j*) monitor the plan;
(*k*) consider wider applicability of solution;
(*l*) restart from (*b*).

A number of specific techniques characterize quality circles, the most prominent of which is *brainstorming.* A subject or problem for which ideas are needed is nominated and the group then suggest as many ideas as possible. As ideas are suggested they are written down on a flip-chart. Other members of the group do not, at this stage, make any comments or criticisms of any suggestions made. At the end of the session the group go through the ideas listed and put similar ideas together and combine overlapping ideas. Typically this technique would be used to generate the initial list of problems or a list of possible solutions to a particular problem.

Also associated with quality circles is the compilation of *most probable error lists.* The project team could be assembled and spend some time producing a list of the most common types of error that occur in requirements specifications. This is then used to identify measures that can reduce the occurrence of each type of error. They might suggest, for instance, that lest cases should be produced at the same time as the requirements specification and that these test cases should be dry run at an inspection. Another result might be a checklist for use when conducting inspections of requirement specifications.

The effectiveness of quality circles

For quality circles to work there must be full support for them at all levels of management. First-line management can feel threatened by them as they might appear to undermine their authority. After all, problem solving is one of their main tasks. The proponents of quality circles see them as a way of giving management more time for planning and development. Any manager will have to devote time to 'fire-fighting', dealing with ad hoe crises, and this can detract from longer term activities that will be able to improve the effectiveness of the organization.

The GQM approach

Approaches to process improvement, such as the quality circles that have been just discussed, may be supported by quantitative measurement techniques. One of these is the GQM (Goal/Question/Metric) approach. This requires that the goal to be achieved is firstly identified. A goal might be to evaluate whether a new programming language allowed developers to be more productive. In order to achieve this goal, certain specific *questions* now need to be identified as needing answers. For example, in order to evaluate the productivity of a new programming language the following questions might need answering:

- how quickly were developers previously writing code?
- how quickly are developers writing code with the new programming language?
- how good was the quality the software previously?
- how good is the quality of the software produced using the new programming language?

For each question, a number of *metrics* will need to be identified as needing collection in order to answer the question. For instance, to find out how quickly developers have been writing code, development effort and the size of development tasks in terms of the number of function points to be produced might be needed.

CONCEPTUAL SHORT QUESTIONS WITH ANSWERS

1. Which CASE tool can be used for generating test cases?

Ans. 'Test case generators' can be used for generating test cases. For example, astra quick test is a test case generator tool which captures business processes in the visual map and generates data driven tests automatically.

2. What are the various types of testing strategies?

Ans. (*i*) Analytic Testing (*ii*) Model-Based Testing
(*iii*) Methodical Testing (*iv*) Process-Oriented Testing
(*v*) Dynamic Testing (*vi*) Philosophical Testing

3. What is the first law of system engineering?

Ans. The first law of system engineering state that "No matter where you are in the system life cycle, the system will change it will persist throughout the life cycle.''

4. CMM stands for?

Ans. CMM stands for Capacity Maturity Model.

5. KPA stands for?

Ans. KPA stands for key process area.

6. What are the number of clauses in ISO-9001?

Ans. 20

7. Name the mechanism that can be used to check the correctness of the output of the program for the test cases?

Ans. Test Oracle.

REVIEW QUESTIONS

1. What do you understand by the term quality? Define software quality and discuss its importance.
2. Write short notes on:
   (*a*) ISO 9126
   (*b*) Product *vs* Process quality management.
3. Define software quality. How ISO 9126 helps in achieving it?
4. Why is the cost of quality failure very high and damaging? Describe the following quality parameters.
   (*a*) Operability
   (*b*) Traceability
   (*c*) Consistency
   (*d*) Completeness.
5. Is SQA an important activity?
6. A software has a login-id and password for security reasons. What static test cases you need to test the functionality of this software?
7. What is statistical quality assurance for software? How it is done?
8. Explain why a high-quality software process should lead to high quality software product. Discuss possible problems with the system of quality management.

CHAPTER 10

STUDY OF ANY SPM SOFTWARE—PROJECT 2000 OR EQUIVALENT

INSIDE THIS CHAPTER

Microsoft Project (or MSP) is a project management software program developed and sold by Microsoft which is designed to assist project managers in developing plans, assigning resources to tasks, tracking progress, managing budgets and analyzing workloads. The first version of Microsoft Project was released for DOS in 1984 by a company working for Microsoft. Microsoft bought all rights to the software in 1985 and released version 2. Version 3 for DOS was released in 1986. Version 4 for DOS was the final DOS version, released in 1986. The first Windows version was released in 1990, and was labelled version 1 for windows. An interesting thing to note is that the first version for DOS introduced the concept of link lines (dependency lines) between tasks in the Gantt chart.

Microsoft Project was the company's third Windows-based application, and within a couple of years of its introduction WinProj was the dominate PC-based project management software. Although branded as a member of the Microsoft office family, it has never been included in any of the office suites. It is available currently in two editions, Standard and Professional.

A Macintosh version was released in July 1991 and development continued through Project 4.0 for Mac in 1993. In 1994, Microsoft stopped development of most of its Mac application and did not offer a new version of Office until 1998, after the creation of the new Microsoft Makintosh Business Unit the year prior. The MacBU never released on updated version of Project, and the prior version from 1993 does not run natively on Mac OS X.

Further versions were released in 1992 (v3), 1994 (v4), 1995, 1998, 2000, 2003 and 2007. There was no Version 2 on either platform; the original design SPEC was augmented with the addition to macro capabilities and the extra work required to support a macro language pushed the development schedule out to early 1992 (Version 3). The application creates critical path schedules, although critical chain and event chain are visualized in a Gantt chart. Additionally, Project can recognize different classes of users can have differing access levels to projects, views, and other data.

Custom objects such as calendars, views, tables, filters and fields are stored in an enterprise global which is shared by all users.

Microsoft Project and Project Server are the cornerstones of the Microsoft Office Enterprise Project Management (EPM) product.

Microsoft has revealed that the next version of Microsoft Project will be featuring the Fluent user interface.

Project creates budgets based on assignment work and resource rates. As resources are assigned to tasks and assignment work estimated, the program calculates the cost equals the work times the rate, which rolls up to the task level and then to any summary tasks and finally to the project level. Resource definitions (people, equipment and materials) can be shared between projects using a shared resource pool. Each resource can have its own calendar, which defines what days and shifts a resource is available. Resource rates are used to calculate resource assignment costs which are rolled up and summarized at the resource level. Each resource can be assigned to multiple tasks in multiple plans and each task can be assigned multiple resources, and the application schedules task work based on the resource availability as defined in the resource calendars. All resources can be defined in an enterprise-wide resource pool.

MS Project presumes additional physical raw materials are always available without limit. Therefore it cannot determine how many finished products can be produced with a given amount of raw materials. This makes MS Project unsuitable for solving problems of available materials constrained production. Additional software is necessary to manage a complex facility that produces physical goods.

10.1 MS PROJECT TUTORIAL – LESSON I: TASKS

To Start a New Project File

You can create a Project from a Template File by choosing File > New from the menu. In the New File dialog box that opens, select the Project Templates tab and select the template that suits your project best and click OK. (You may choose a Blank Project Template and customize it)

Now, you need to define for Microsoft Project the start or ending date for the project. If you fill in a start date, Project schedules your tasks from that date, based on the order in which tasks need to be completed and the availability of resources assigned to work on each task. But, if you have to finish the project by a deadline date, you can enter the finish date too. All intermediate schedules will then be set by Project. Once a new Project page is opened, the Project Information dialog box opens as shown below.

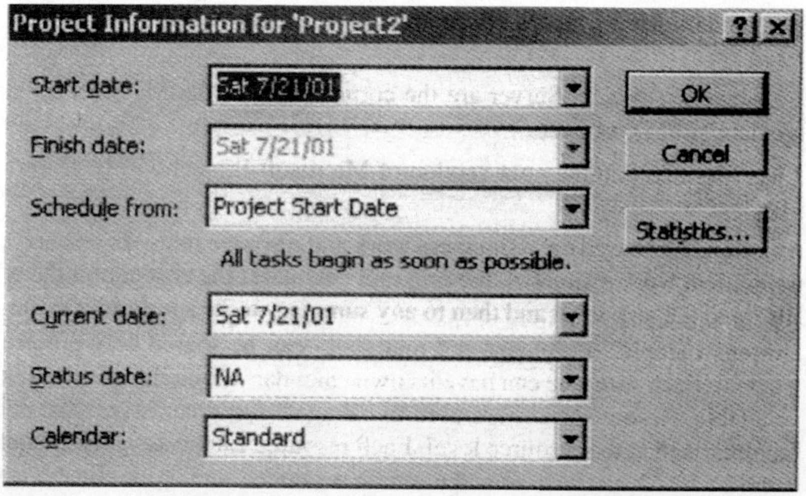

Enter the start date or select an appropriate date by scrolling down the list. Click OK. Project automatically enters a default start time and stores it as part of the dates entered. Click on Save and type in the name of your file in the dialog box that opens.

To Record Properties

1. Choose File, Properties from the menu to display the Properties dialog box.
2. Select Summary tab and fill in information in required fields, (project title, project manager, company name, project goals).
3. Click OK.

You can set default values for Start Time, Hours per day/week by selecting Options from the Tools menu.

Views

Views allow you to examine your project from different angles based on what information you want displayed at any given time. You can use a combination of views in the same window at the same time.

Project Views are categorized into two types:

- Task Views (5 types)
- Resource Views (3 types)

To find about more about each view, use the scroll buttons at the top and bottom of the View bar to scroll down through available views. The Project worksheet is located in the main part of the application window and displays different information depending on the view you choose. The default view is the Gantt Chart View. In Microsoft Project, graphical bars are paired with a spreadsheet to give the Gantt Chart View. The next step is to create an outline of the work that has to be done on the project. Goals of any project need to be defined in terms of tasks.

There are four major types of tasks:

1. *Summary tasks* – contain subtasks and their related properties
2. *Subtasks* – are smaller tasks that are a part of a summary task
3. *Recurring tasks* – are tasks that occur at regular intervals
4. *Milestones* – are tasks that are set to zero duration and are like interim goals in the project

Entering Tasks and Assigning Task Duration

Click in the first cell and type the task name. Press enter to move to the next row. By default, estimated duration of each task is a day. But, there are very few tasks that can be completed in a day's time. You can enter the task duration as and when you decide upon a suitable estimate.

Double-clicking a task or clicking on the Task Information button on the standard toolbar opens the Task Information box. You can fill in more detailed descriptions of tasks and attach documents related to the task in the options available in this box.

To enter a milestone, enter the name and set its duration to zero. Project represents it as a diamond shape instead of a bar in the Gantt Chart.

To copy tasks and their contents, click on the task ID number at the left of the task and copy and paste as usual.

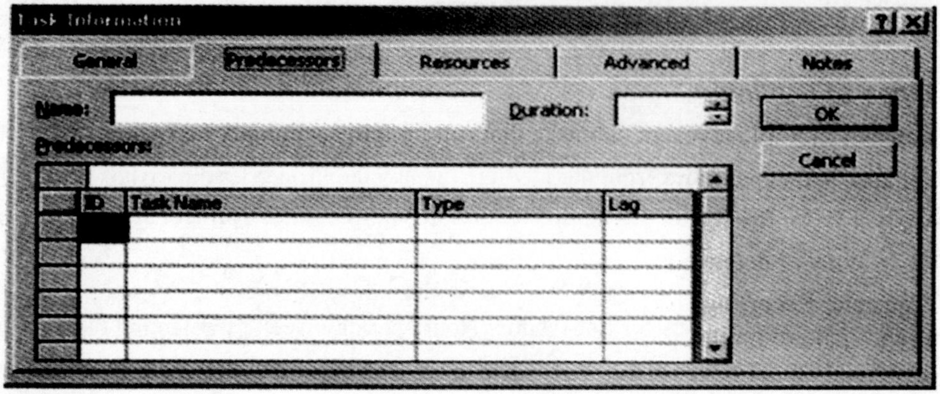

Outlining Tasks

Once the summary tasks have been entered in a task table, you will need to insert subtasks in the blank rows and indent them under the summary task. This is accomplished with the help of the outlining tool.

Outlining is already active when you launch a project and its tools are found at the left end of the Formatting bar. To enter a subtask, enter the task in a blank cell in the Task Name column and click the Indent button on the Outlining toolbar. The Show feature in this toolbar is drop down tool that gives you an option of different Outline levels.

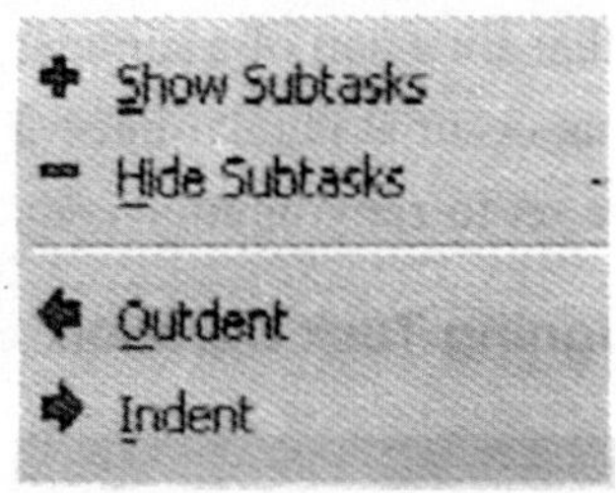

A summary task is outdented to the left cell border, is bold and has a Collapse (-) (Hide subtasks) button in front of it and its respective subtasks are indented with respect to it.

Advantages of Outlining

- It creates multiple levels of subtasks that roll up into a summary task
- Collapse and expand summary tasks when necessary
- Apply a Work Breakdown structure
- Move, copy or delete entire groups of tasks

You can enter Recurring tasks by clicking on Insert > Recurring task and filling in the duration and recurrence pattern for the task. Any action you perform on a summary task – deleting it, moving or copying it apply to its subtasks too. You can also create a custom outline code by choosing Tools > Customize > Fields and select the outline code in the Type down list and click Define Outline Code and Help. This will take you to online help for outline code creation.

Links

Tasks are usually scheduled to start as soon as possible *i.e.* the first working day after the project start date. The remaining schedule can be defined in a base calendar to which Project is linked. Project supplies three calendars and any one of them can be used. The Standard calendar, 24 Hours calendar and the Night Shift calendar schedule tasks on different working days and hours. The Standard calendar is usually used as a base calendar. The duration of any task can be seen in the form of gray bars of varying length on the Gantt chart in the timeline section.

Sometimes, even after designing a schedule many unforeseen changes can affect the completion of tasks midway. This could lead to a disruption to completion of other related tasks. To solve this problem, Project allows you to link tasks in various Ways. By linking tasks Project adjusts the schedule whenever there are changes that affect duration of other tasks.

A task that needs to be completed before (painting the walls of a home before moving the furniture) are called predecessor task and the linked tasks are its successors.

Tasks can be linked in four ways:

(*a*) ***To Link Tasks in FS Dependency***

1. Select the tasks you want to link. Tasks with a lower ID number are taken as predecessors, but if you want to set the order of tasks yourself, select tasks by holding Control while you click the tasks in order.
2. Click Link Tasks button on Standard Tool bar or choose Edit > Link tasks or Control+F2.

Links	Code	Description	Gantt chart view
Finish-Start	F-S	Predecessor finishes and the other starts	
Start-Finish	S-F	Task begins at the same time as its predecessor	
Finish-Finish	F-F	Both tasks finish at the same time	
Start-Start	S-S	Start of the predecessor determines when the other starts	

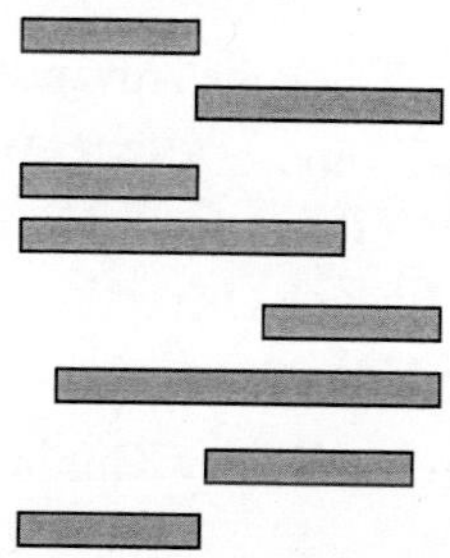

Summary tasks can be linked to other summary tasks or to subtasks between summary groups. Subtasks can be linked to each other too. To unlink tasks, select tasks and click the Unlink Task in the Standard Toolbar. Do not attempt to remove a link by deleting a cell in the predecessor's column and pressing delete as this will delete the entire task.

(*b*) ***Delayed and Overlapped Links***

Sometimes tasks may be dependent on other factors causing a delay time or an overlap time, necessitating a lag or lead-time to be incorporated in a link. Lag and lead-time can be entered as units of time or as a percentage of the duration of the predecessor.

1. Double-click a task and open the Task Information dialog box.
2. Click the predecessor tab.
3. Select the lag/lead field.
4. Enter a number or a percentage in the cell.
5. Click OK.

The Split task button splits tasks that may be completed in parks at different times with breaks in their duration times. The Auto Link feature is enabled by default and keeps linked tasks intact when they are linked in the FS dependency. It can be disabled in the Options dialog box if necessary.

Constraints

Certain tasks need to be completed within a certain date. Intermediate deadlines may need to be specified. By assigning constraints to a task you can account for scheduling problems. There are about 8 types of constraints and they come under the flexible or inflexible category.

(*a*) ***To apply a constraint***

1. Open the Task Information dialog box.

2. Click the Advanced tab and open the Constraint type list by clicking on the drop-down arrow and select it.

3. Select a date for the Constraint and click OK.

Flexible constraints (demarcated by a red dot in Microsoft Project 2000) restrict scheduling to a great extent whereas flexible constraints (blue dot) allow Project to calculate the schedule and make appropriate adjustments based on the constraint applied.

Inflexible constraints can cause conflicts between successive and preceding tasks at times and you may need to remove such a constraint.

(*b*) ***To remove a constraint and apply an ASAP/ALAP constraint***

1. Select the tasks by holding Control and clicking on them in the order you want.

2. Click Task Information button on the Standard Toolbar and the Multiple task dialog box opens.

3. Click the Advanced tab and click the drop down arrow to open the constraint type list.

4. Choose ASAP if the project has been scheduled from the Start date and ALAP (As late as possible) if it has been scheduled from the Finish date.

5. Click OK.

Or

Select Tools > Options > Schedule and clear the 'Tasks will always honor their constraint dates' checkbox.

Or

Set deadline date instead of adopting an inflexible constraint. Setting a task deadline does not affect the task schedule in any way. If a deadline passes without the task being completed project indicates it in the Indicator column (a downward pointing arrow in the Gantt chart view).

10.2 MS PROJECT TUTORIAL – LESSON 2

In this lesson we will learn to successfully schedule all the resources required to complete a project – people, equipment, supplies and materials. We will also learn to include costs of various kinds and assign resources to tasks.

There are three main parts to this lesson:

1. Resources (*a*) Entering Resource Information; (*b*) Resource lists; (*c*) Resource information dialog box; (*d*) Assigning a base calendar other than the project calendar to the resource; (*e*) Editing a resource calendar

2. Assignments (*a*) Assigning a resource to a task; (*b*) Task types; (*c*) Changing the default task type; (*d*) Assigning work resources to tasks; (*e*) Assigning work resources in the task form; (*f*) Assigning material resources to tasks; (*g*) Assigning material resources to a task in Gantt chart view

3. Costs (*a*) Defining and controlling costs; (*b*) Accrual methods; (*c*) Entering cost information; (*d*) To enter costs for a resource.

Resources

Once you determine that you need to include resources into your project you will need to answer the following questions:

- What kind of resources do you need?
- How many of each resource do you need?
- Where will you get these resources?
- How do you determine what your project is going to cost?

Resources are of two types – work resources and material resources. Work resources complete tasks by expending time on them. They are usually people and equipment that have been assigned to work on the project. Material resources are supplies and stocks that are needed to complete a project. A new feature in Microsoft Project 2000 is that it allows you to track material resources and assign them to tasks.

When a set of resources is available for working they are called the resource pool. After you determine the number of resources that you need, you need to establish the time and availability of each resource. In the case of work resources, the amount of time that they can work for, be it in hours, days or months, or years and the amount (units of measurement) of material resources need to be specifically defined.

The next step is to assign these resources to their respective tasks. When you allocate a resource's time to work on a task you are assigning resources. Once this is done project can recalculate the schedule to accommodate the working times of the assigned resources. Vacations and nonworking times can be identified and included for each resource. It goes one step ahead and tells you when you have overallocated a resource, *i.e.* when you have assigned a resource to multiple tasks in the same time period or when a resource is assigned to do more work than it can complete in a certain time.

(*a*) ***Entering resource information in project***

The assign resources dialog box is used to create list of names in the resource pool.

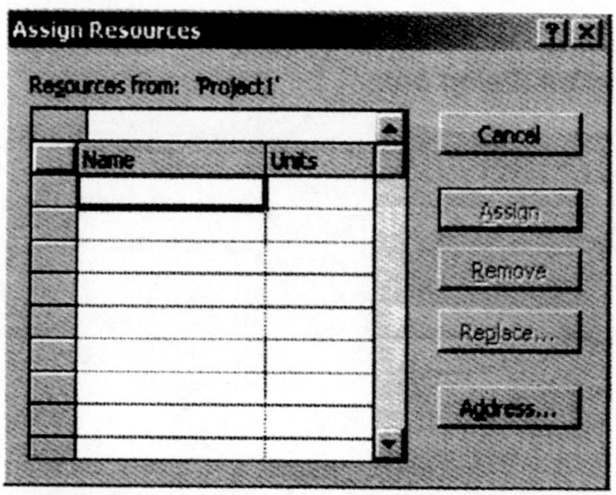

(*b*) ***To enter resource lists***

1. Click the Assign Resources button on the Standard Toolbar or Tools > Resources > Assign resources.

2. Select a cell in the name column and type a response name. Press Enter. (You can also click adresses and add addresses from Microsoft Outlook Express)

3. Repeat step 2 until you enter all the resource names.

4. Click OK.

Resource names cannot contain slash (/), brackets [] and commas (,).

Another way of defining your resource list is through the Resource Sheet View.

1. Display Resource Sheet View by choosing view > Resource Sheet or click the Resource Sheet icon on the Viewbar.

2. Enter your information. (To enter material resource use the Material Label field)

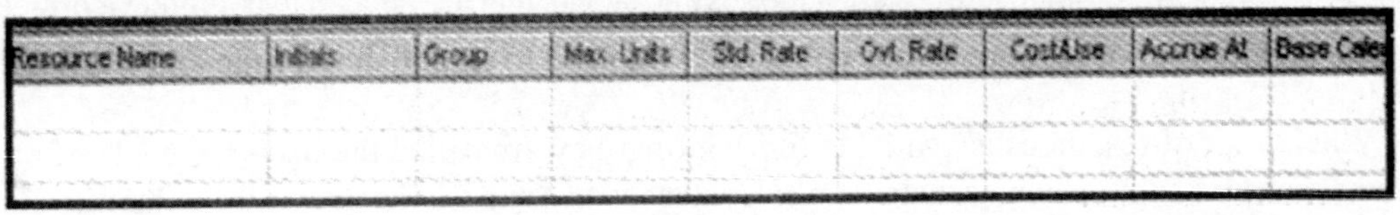

The Max Unit's column contains the maximum capacity for which a work resource is available to accomplish any tasks in a certain time period. The default format is the percentage format. For example, if the assigned units are 100% this will imply that this resource will work for all the eight hours in a normal working day. If the assigned units are 50%, then this resource will work for four hours a day. In the case of a resource pool consisting of 5 workers, their maximum units will be 500% or a decimal value of 5.

You can change the default format to decimals using Tools > Options > Schedule tab > Show Assignment Units as. Material resources don't have Max Units. Their resource units show their consumption rates that could be fixed or variable. The other three columns in the Resource Sheet View are concerned with costs and we shall take a look at them a little later in this lesson.

(*c*) ***Resource information dialog box***

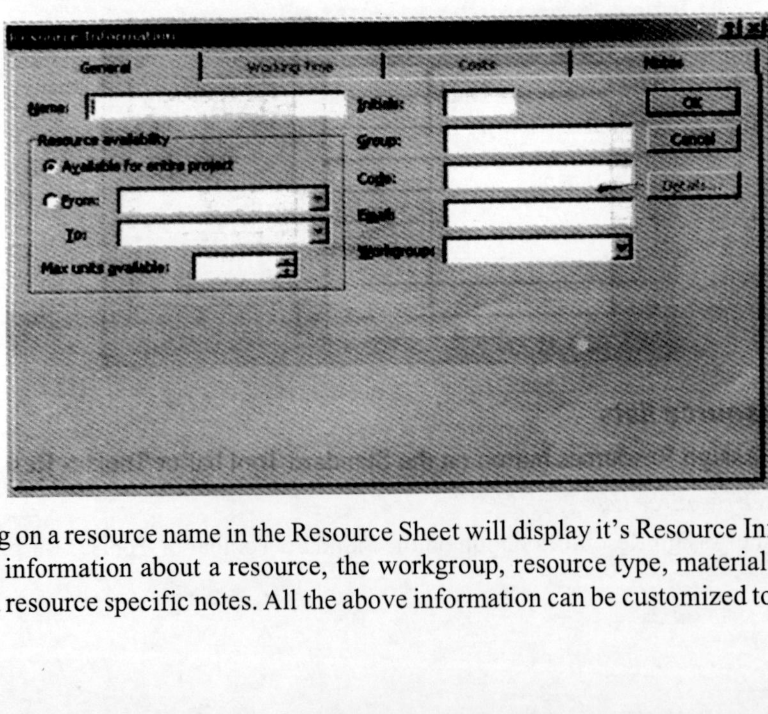

Clicking on a resource name in the Resource Sheet will display it's Resource Information dialog box that gives information about a resource, the workgroup, resource type, material labels, working time, costs and resource specific notes. All the above information can be customized to suit your needs.

With the Working Time tab in this box, you can assign a different base calendar to each resource and change the working times of that resource as and when required.

(*d*) ***To assign a base calendar other than the project calendar to the resource***

In the Resource Information dialog box switch to resource view and double-click resource and click on the Working Time tab to make your changes.

(*e*) ***To edit a resource calendar***

Tools > Change Working Times > Select the calendar you want to edit.

Workgroup defines how you intend to communicate with the resources in the project. You can use Microsoft Project Central to communicate using the Web.

Assignments

(*a*) ***To assign a resource to a task:***

1. Open the Gantt Chart View and the Assign Resources Dialog box.
2. In the Entry Table select the tasks for which you want to assign resources.
3. In the Assign Resources Dialog box, select the resource (resources) you want to assign.
4. Click the Assign button.

You can change or assign more resources to tasks using the Assign Resources Dialog box. Setting up a project file, assigning resources to tasks and changing resource assignments during the course of the project are fine, but how do you incorporate unpredictable factors that can affect task duration and work assignments.

How does Project 2000 help you with this?

When you assign resources to tasks, Project takes the task's duration and multiplies it by the specific percentage or number of units of the resource to arrive at the hours of work that will be done in the task. Work is amount of time or effort that a work resource expends on a task, *i.e.* 16 hours of shipping, 5 days of training, and so on.

Duration is the amount of clock time between the beginning and ending of a task. So, Work = Duration * Units (W = D * U).

For a particular task, lets say, in developing a certain training program, the duration is 5 days and 3 resources have been assigned to work at 100% units. Considering that the normal working hours in a day (by default) are 8 hours, each one of these resources will work for 8 hours on all 5 days. Now, if two part-timers with units of 50% were assigned to the project, then, they would work for 4 hours a day. This will result in the work being completed earlier than the stipulated 5 days and the amount of work (time) expended by the first 3 resources will decrease. In such a case, project will need to recalculate schedules from its default settings. Given the choice between duration, work and units, project will choose to change duration. If the duration happens to be fixed, then it will change work before it changes units.

You can assign a greater control over project's calculations by defining the task type as described in the next section.

(*b*) ***Task Type***

Project defines three task types that help it recalculate variables when resources are assigned to it:

- Fixed Units
- Fixed Work
- Fixed Duration

Each task type definition is based on the variable that is kept constant. For the Fixed Units task type, project will recalculate work or duration as whenever necessary, but work units will remain fixed. Similarly, in Fixed Duration tasks, if the units are changed, work will be recalculated accordingly. When using Project 2000 to manage a project, a project manager will need to analyze each task and then choose the suitable task type. The default task type is the Fixed Units task.

(*c*) ***To change the default task type***

1. Tools > Options > Options Dialog box.
2. Schedule tab > Fixed Duration or Fixed Work from the Default Task Type drop-down list.
3. Click OK.

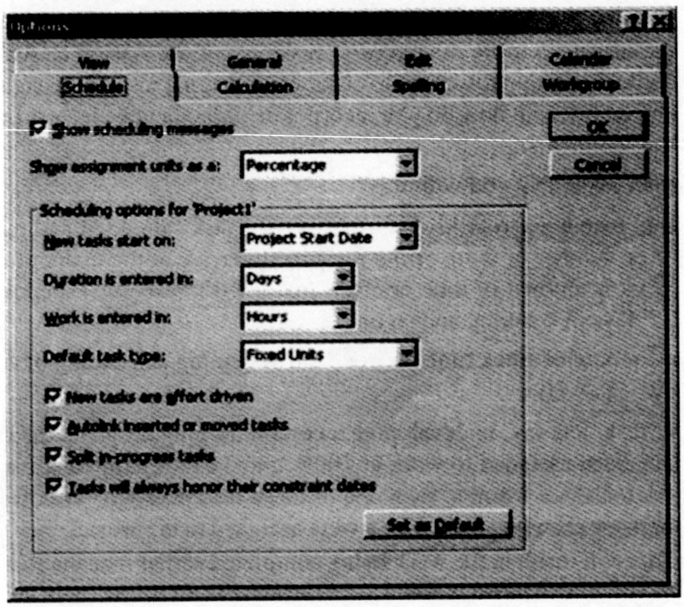

Also, the default settings for Tasks are effort-driven. The task's duration is completely dependent on the resources you assign to it. If you change the number of resources after the initial assignment, Project will recalculate duration but leave work alone. You can use Tools > Options > Options Dialog box > Schedule, to turn off "New Tasks are Effort-driven" check box.

(*d*) ***Assigning work resources to Tasks***

The Task Form can display resource assignment details like Units and work. To display the Task Form along with the Gantt chart click view > Gantt chart > Split the view by choosing Windows > Split. The Task Form is automatically displayed in the lower half pane. Right-clicking there will display the short-cut menu.

(*e*) ***To assign work resources in the Task Form***

1. Select the task in the Gantt Chart View.
2. In the Task Form, select a resource from the Resource name drop-down list.
3. In the units and work columns, use spin box controls to enter units, work or both.
4. Repeat steps 2 and 3 until all the resources are chosen for the selected task and click OK in the Task Form to assign the resource.

You can also assign resources using the Assign Resources Dialog box described earlier.

(*f*) ***Assigning Material Resources to Tasks***

When you assign a material resource, you specify the way the material is consumed. Choosing 'variable material consumption' changed the task duration. (For example, 6 gallons/hour or 4 pk/h) Fixed material consumption will not change task duration.

(*g*) ***To assign material resources to a task in Gantt chart view***

1. Select the tasks in the Gantt chart pane.
2. Select the Resource Name field in the task form.
3. Select the resource name from the drop-down list.
4. Select the units field and enter the units you want to assign. If you leave this field blank, Project will set default of 1.
5. Project will append the Material Label to your decimal entry.
6. Work Field Entries: For fixed consumption rates, the value in the Work Field and Units Field will be the same, whereas for variable consumption rates, the Work Field value will be the Units Field value multiplied by the task duration.
7. After all the resource assignments are made for the task, Click OK.

If you intend removing an assignment from a task, select the task and then the resource name you want to remove and press the Delete key. The Effort Driven Task Field appears on the Task form and in the Task Information Dialog box as a checkbox that is filled by default in Microsoft Project.

Costs

(*a*) ***Defining and controlling costs***

Project multiplies the number of hours a resource works and the hourly cost rate to give you the resource cost rate. For multiple resources the individual resource cost is added. The number of units used reflects material costs.

The Standard Rate or the Rate of pay per time unit for regular working hours or the cost per unit of a material resource and the Overtime Rate defined for overtime hours are the two rate values that have to be filled in the Resource Sheet view. Fixed costs and the resource costs add up to the total costs.

(*b*) ***Accrual methods***

The accrual methods describe the ways in which Project calculates costs. Sometimes, you may have to pay a resource in full before it starts work, then the accrual method adopted by Project is 'Start'. Likewise, when payments have to be made only after completion of the task, the 'End' method is used. For tasks that are partially completed, estimated costs are prorated, *i.e.* if 20% of the work is completed, the calculated cost at that time will be 20% of the estimated cost. This is where accrual methods play an important role. They don't have any effect on the final cost of the project.

(*c*) ***Entering cost information***

On the costs tab of the Resources Information Dialog box you can record up to 5 different rates for a selected resource that is assigned to 5 different tasks. You can also enter separate rates like the Standard Rate, Overtime Rate and a Per Use Cost.

(*d*) ***To enter costs for a resource***

1. Open Resource Information Dialog box and click costs tab.
2. Effective date of the cost may be the same as the project start date. If not so, enter a date or select a date from the drop-down calendar.

3. Enter the Standard Rate. For work resources the default is $/h, but you can type in /y, /m as necessary. Material resources use the Material Label assigned to the resource (gallons or boxes).

4. Enter an Overtime rate (enter it even if it is the same as the Standard Rate).

5. Enter any additional Per Use Costs for a resource (costs for shipping and so on).

6. If you know that rates may change, click in the second row, type an effective date and enter new standard, overtime and Cost Per Use rates. You can fill in the actual rates or enter them as a % decrease or increase from the rates in the first row.

7. If resources have different accrual methods, select your choice from the cost accrual drop-down list.

Project supports fixed and variable costs. Fixed costs are those that do not change over the life of the project. To enter fixed costs choose View > Table > Cost and open the Cost table. Select the Fixed Cost field for the task and enter the amount. In addition to the costs listed above, you can also have a fixed resource cost. This is when a stipulated amount has to be paid to a resource irrespective of the work done by it. To include these costs, split the Gantt Chart View and activate the bottom resource pane and follow these steps:

1. Choose Format > Details > Resource Costs to display cost details.
2. Select the task and set its Task Type to fixed duration.
3. Enter the Resource name and type % in the units field.
4. Click OK. (This tells project not to calculate work)
5. Enter values in the Cost Field.
6. Click OK.

Additional changes to resources assignments can be made using the Task or Resource Usage View and the Assignment Information Dialog box.

10.3 MS PROJECT TUTORIAL – LESSON 3

Now that we have seen how resources are assigned to tasks and costs are assigned to resources, let us take a look at handling contours and overallocated resources. We will also learn to fine-tune our project and prepare it for publication.

1. Contours
2. Overallocated Resources
3. Fine Tuning and Preparing the Project for Publication.

Contours

Resources are usually human beings and therefore they bring a lot of unpredictability with them - dates can change, resources may be assigned too much of work or they may have to leave in the middle of a project. We have to leave room for changes in assignments and work distributions. Project allows you to arrange for resources to start or finish work at different dates, interrupt assignments to allow for work on other tasks and also assign overtime work if necessary – in other words a contouring of assignments is possible. The Task Usage View and the Resource Usage Views offer such possibilities.

What does the term contour mean in Project? It is the shape of the distribution of work within an assignment. Usually, if a resource is assigned 100% to a 40-hour task with no predecessors, the task will begin immediately with 8 hours of work/day for 5 days. This is the default flat work contour where each unit's work is spread evenly across the duration of the task.

Sometimes, you may need to change the distribution of an assignment's work by applying a predefined contour or creating a custom contour for it. Project 2000 includes 8 predefined contours as listed below:

- **Flat:** The default contour with an even distribution of work.
- **Back-loaded:** Peak activity happens at the end of the project.
- **Front-loaded:** Peak activity is at the beginning of the project.
- **Double peak:** A project that has two major periods of peak activity.
- **Early peak:** The same as the front-loaded but with a ramp up to peak activity
- **Late peak:** A back-loaded contour with a ramp.
- **Bell:** Single peak in the middle of the project
- **Turtle:** A bell with a ramp up and ramp down.

To change the work contour for an assignment in either Task Usage View or the Resource Usage View, double-click an assignment to open the Assignment Information Dialog box and choose a contour from the Work Contour drop-down list.

Contours interact with task and resource settings as you manage and adjust assignments during the life of the project. The contour shape is preserved with the assignment when you move the task or when the project schedule changes.

When you manually change the work assignments in the Resource Usage View or the Task Usage View, the result is an edited contour. Whenever you edit a contour, project distributes the work duration and recalculates the units.

Overallocated Resources

A resource is overallocated when the total of its time phased work (work that is distributed over specific periods in the project) exceeds the resource's maximum units. If a resource is already scheduled to be working on a different task, the resource may become overallocated. Sometimes, project also accidentally includes overallocated resources. In every resource view, overallocated resources are formatted in red. In the Resource Sheet, project displays a caution icon in the indicator column of overallocated resources. Hover over this icon and project will display a tip to correct this overallocation.

Locating the overallocations

Combine the Resource Graph with the Resource Form to view assignments and resource information with the resource's schedule. Choose View > Resource Graph and Window > split to open all the above mentioned views. Right clicking on the graph will open the shortcut menu.

Choose overallocation and ways to resolve overallocations

You can choose to:

- To delay a task
- Split a task
- Assign additional or different resources

Check the Remaining Availability option in the shortcut menu to check on times available for that resource. To assign additional or different resources display the Resource Usage View and choose Group from the Group drop-down list to arrange resources in groups. Project 2000 offers a levelling feature that can resolve resource allocations by splitting or delaying tasks. It uses factors like:

- Available time
- Task priority, dependencies and constraints
- Task ID
- Scheduling dates

To apply the levelling feature

1. Select any task.
2. Choose Tools > Resource Levelling. This will open the Resource Leveling Dialog box.
3. Select Level From/To option and use the drop-down calendar to select beginning and ending dates between which project should search for overallocations. After it identifies overallocations, it delays tasks based on the levelling order selected.

Tasks with constrains like Must Finish On, As Soon as Possible or As Late as Possible are left alone, so also tasks that have already begun.

Three levelling orders are:

1. ID: Higher ID tasks are delayed (It would be useful to enter tasks in order of importance).
2. Standard: This is the Default levelling order.
   - Tasks without successors are delayed first.
   - Tasks that are not on a critical path (most important work path of the project) are delayed first.
   - Lower-priority and non-constrained tasks are delayed first.
   - Tasks with later start dates are allowed to begin later.
3. Priority/Standard Levelling Order: The only factor of importance is the priority of tasks. Task priority is set to any value between 0 and 1000 on the General Tab of the Task Information Form. (If you don't want Project to delay a task set its priority to 1000 and use this levelling order).

To level your resources on your own

Display the Resource Allocation View with assignment delay and task delay fields visible.

1. Choose View > More Views > Resource Allocation.
2. Click in the Gantt Chart in the lower pane.
3. Choose View > Table > More Tables and the More Tables Dialog box opens.
4. Select Delay.
5. Click Apply.
6. In the Resource Usage Table in the upper pane, right-click on Work Column.
7. Choose Insert Column from the shortcut menu.
8. Select Assignment Delay Column in the Column Definition Dialog box.
9. Click OK.

To add delay to a task

1. Select any assignment for the task you want to delay.
2. In the leveling delay field in the lower pane, enter the duration of the delay.

To delay an assignment

When you delay an assignment, you are effectively delaying the time when the resource begins working on an assignment after the task's start date has been set.

1. Select the assignment that you want to delay in the Resource Usage Table.
2. Enter the duration for the delay in the Assignment Delay Field.

Fine Tuning and Preparing the Project for Publication

So far you have dealt with tasks, resources, assignments and costs in detail. Your project design is almost complete, leveled and ready for presentation and you are waiting for the go ahead to implement the project. Just before you make that all-important presentation it would be useful to step back and evaluate your work. To zoom in on the overall quality and finesse of your project, Microsoft Project 2000 provides you with a number of tools.

You can:

- Display a summary of all the important statistics using the Summary table display
- Use filters to zero in on the information you want to view
- Sort and group information
- Identify and implement strategies to reduce duration and costs

The Project Statistics Dialog box provides summary information about your project.

At the top of this box you will see the current start and finish dates. When you initially save a project, Project's Planning Wizard prompts you to save the project with or without a baseline. Now, what is a baseline? It is a project plan with original estimates for tasks, resources, assignments and costs. With it, you can compare task, resource and other updates as the project work is completed. These variances help identify potential problems in the project.

In the Project Statistics Dialog box, Actual row gives you current dates while the variance row displays the variance between baseline and scheduled dates. This Dialog box gives information about duration and costs too.

Changing Timescales

The default timescale in the Gantt chart view are weeks and days. If you choose to get a broader picture of the project you can zoom out the timescale to months and weeks. Clicking on the Zoom Out button on the Standard Tool bar will allow you to select the set of time units of your choice.

Collapsing the task outline

Display any view showing the task lists.

1. Click the Show Drop-down arrow on the Formatting Tool bar and select an outline level to view. These outline levels display subtasks indented in various ways.
2. To get back to normal viewing of all the tasks, click the Show Arrow and select All Subtasks.

Filtering

Filtering allows you to impose certain conditions on the way in which you want to display your information at any time. You can choose to hide information that you don't want to view or highlight the information that is important to you. The Auto Filters are most suitable for this need.

To apply an Auto Filter

1. Choose Project > Filtered For > Auto Filter.
2. Click on one of the drop-down lists to view the various headings and select the one you want.
3. Once the table is filtered the column headings and filter drop-down arrow turns blue.
4. To include additional filter conditions use the drop-down list to select other headings.

5. To turn off the Auto Filter select Project > Filtered for > Click Auto Filter to turn the feature off.

You can also filter tasks using the predefined filters that project provides you with. You can use them to filter tasks like milestones, summary tasks, critical tasks, resource-specific tasks and incomplete tasks.

Sorting and Grouping

Changing the sort order of your data organizes them in a clearer manner. You can sort by ascending (default order), cost, priority, Start and Finish dates. For resource lists, standard sorts are available for name, ID and costs. You can apply multilevel sorting using Project > Sort > Sort by to open the Sort Dialog box.

In addition to sorting resources or tasks you can categorize them into groups. For example, your resources may belong to different categories like management, support or training.

To create a custom group:

1. Display the Gantt Chart View or Resource Sheet View and Select View > Table.
2. Select Project > Group by and choose from the cascading menu of existing group definitions.
3. To add a new group definition choose More Groups to open the More Groups dialog box and create a group of your choice.

Viewing Costs

The Summary Table gives you a quick look at cost and work.

1. Display the Gantt Chart or other list views.
2. Choose View > Table > Summary. Columns for duration, start, finish, cost, work and percentage of completion are displayed.

Reducing Costs

After opening the Summary Table and viewing the costs you can open the Gantt Chart View along with the Task Form and select a task in the Gantt Chart. Right-click on the Task Form and select Resource Cost settings to take a look at the complete list of costing fields. As we learnt in Lesson 2, resource and fixed costs add up to total costs. Among these two cost types, the resource costs are usually the higher costs in any project. Reducing resource costs (without compromising on quality), keeping a watchful eye on any task that may go over the budget and doing away with tasks that are quite unnecessary are sure ways to reduce project costs.

The all-important sets of tasks in any project – those that are vital to the completion of a project are the critical ones and their path or sequence is referred to as the critical path. These tasks may not allow interruptions but it makes sense to look at ways to shorten this critical path.

To view critical tasks

In the More Filters Dialog box, select critical from the list of filters and click on Highlight. All critical tasks will appear in a different color from the other tasks.

To reduce the critical path

1. Overlap tasks
2. Schedule overtime
3. Break large tasks into smaller ones

4. Identify and correct any errors
5. Reduce the scope – complete all planned tasks within a timeframe.

10.4 MS PROJECT TUTORIAL – LESSON 4

In this lesson we will learn to customize how information in project is displayed, format and modify views (refer lesson 1 for a recap on views) and print the finalized schedule. We will see how project files can be saved as web pages and create hyperlinks to navigate to other files and web pages.

1. Formatting Views
2. Publishing Projects on the Web or Intranet
3. Hyperlinks

Formatting Views

There are dozens of predefined views available in Project 2000 that can be customized to suit your needs. Selected text and timescales can be formatted, one can zoom in and out to get a closer look at the view shown. Formatting options help customize the way you want your information displayed.

Formatting Text

You can use the Formatting toolbar buttons or the Format Menu options to format selected text. The Format Painter on the standard toolbar helps you copy the formatting style that you applied to any one section of text elsewhere.

1. Select the cell that you want to format.
2. Choose Format, Font to display the Font Dialog box. This allows you to assign font attributes to selected text.
3. Select a new Font type and make the necessary style, size and color changes and click OK.

Formatting Timescales

A timescale area on a Gantt Chart displays two time unit levels:

- Major Units scale
- Minor Units scale

The time units, dates can be formatted to fit into your viewing needs. To format Timescale area:

1. In the Gantt Chart View, choose Format, Timescale to open the Timescale Dialog box.
2. In the Major and Minor scale sections, select the units of time you want.
3. Each set of units has a Label and an Align box. Select the desired values and click OK.

If you wish to adjust the appearance of the nonworking time, click the nonworking time tab of the Timescale Dialog box and choose your options here.

Formatting Gridlines

When information is displayed across rows and columns, the appropriate placement of gridlines is very useful in obtaining a good view. Format – Gridlines will open the Gridlines dialog box. The different lines for Gantt Chart Rows are displayed in a row.

1. Select the line you want to change from the list.
2. In the normal section of the dialog box, select the type and pattern for the gridline.
3. To get rid of an existing gridline, select it from the list and check the blank area in the Type drop-down list.

4. Select an interval, line type and color.

5. In the At Interval section, choose None to display all lines of the selected type or choose an interval to display lines only at that interval.

6. Click OK.

In addition to the above mentioned formatting methods you can format tables, fields and other views and modify them accordingly. The Drawing Tools on the Drawing Toolbar can create objects and buttons that format existing objects in interesting ways.

Okay, budget and schedule optimizations complete, your project is ready for printing and distributing among workgroup members and stakeholders. Views and Reports of various kinds from Project 2000 can be printed easily.

Let us explore printing options in a little detail:

Printing Views

Before printing a view, it is important to select the type of view, do the necessary formatting and change the page setup if necessary. In order to change the print settings for an active view, File, Page Setup will display the Page setup dialog box.

This box gives you information on margins, page headers/footers etc. The Header/Footer tabs have 7 buttons and two drop-down lists that can be used to format text, insert graphics into the header or footer.

Using File, Print Preview to display your View and choose Print to print it. The Print Dialog box shows you choices for printing the current screen view.

Project 2000 allows you to set the starting page number for printouts.

To print Gantt Charts

The page Setup Dialog box gives you a few options regarding printing a Gantt Chart View.

1. You can choose to print all sheet columns of the Gantt Chart Table, regardless of whether they are completely visible on the screen or you may print specified number of columns on all pages.

2. You can print the task notes on a separate page.

3. Fit the Timescale to extend all the way to the right margin.

Printing Reports

With Project 2000, you can print reports about every aspect of your project – Overview reports, Top Level Tasks Report, Critical Tasks Report, Milestones Report, Working Days Report, Cost Reports, Assignment reports and workload reports.

As with printed views, you can change reports using tables and filters.

To print a Report

1. Choose View – Reports to open the Report Dialog box and double-click the category of report that you want to print.

2. Double-click a report and project opens it in the print preview window.

3. Click Print.

4. If you choose to hide a few report details, once you click on the report in the Report Dialog box, click the Edit button. The Report Definition Dialog box appears and select the Details tab.

5. Unclick the check mark in the check box next to the Details option you choose to hide.

6. Click OK.

To create a new Report:

1. Select View, Reports.

2. Double-click the Custom category to open the Custom Reports Dialog box. It gives you a few options – "New – Creates a completely new report" Copy – Creates a copy of the selected report. Organizer – Shares reports in Project 2000.

3. Type a name for the report created.

4. Customize it using the Organizer button (Opens the Organizer Dialog box)

Publishing Projects on the Web or Intranet

As with all Microsoft Office applications, Project 2000 allows you to save your Project as an HTML document that can be viewed on any browser such as Microsoft Internet Explorer or Netscape Navigator.

With Microsoft Project 2000, you can do the following:

1. Use a hyperlink (a place on a Web page that navigates you to another location).

2. Select a specific task, resource and assignment using an Import/Export map and publish it in your HTML document.

3. Customize your HTML document to include graphics.

4. Track progress by communicating among team members via your corporate Intranet.

To save a project file as an HTML document:

1. Choose Menu, File and Save as a Web Page.

2. Click Save in the Save as Dialog box.

3. Project displays the Export Mapping Dialog box that allows you to select the map on which to base your document. This map instructs Project about the type of Data saved in the document. Project comes with 12 predefined map files, but by default it saves your HTML file with the extension -. html.

4. From the Selective Data option on the Export Mapping Dialog box, select the Task List with Embedded Assignment Rows map and click Save.

To view this saved document in a web browser, open any browser and select File, open and Browse to navigate to the HTML file that you want to view.

In addition to the 12 maps defined in Project, to create your own Export Map do the following:

1. From Menu, File choose Save as Web page to open the Save as Dialog box and select the filename and location for your HTML file and click Save.

2. From the Export Mapping Dialog box, click New Map. Define Import/Export Dialog box will be displayed.

3. Select Critical Tasks in the Name type.

4. In the Data to Import/Export section, select the tasks check box.

5. Select the Task Mapping tab to display mapping options for your project's task data in the Define Import/Export Map Dialog box.

6. In the Destination HTML Table Title field, type the text that you would like to appear above the task table. (Critical Tasks in this case)

7. From the Export Filter drop-down list, select Critical Filter.

8. Click the Base on Table, button and select Base Table for field to display the Field Mapping Dialog box.

9. Select the schedule tab and click OK to include all the fields into the Task Mapping tab.

10. To display the total work for a task select Total Slack row and click on the Insert Row button. In the From: Microsoft Project Field columns select the Work Field from the drop-down list.

11. In the To: HTML Table field column, change the text Scheduled – Work to Task Work Estimate. This text will become the new column header for the Work Field in the HTML document.

12. Click OK. The Export Mapping Dialog box will be displayed. Click on Save and open any browser to view the critical task path.

To display Resource Assignments in your HTML document

Follow all the steps listed earlier, in the Define Import/Export map dialog box, check the Include Assignment Rows in the In Output option in the HTML options section and click OK and save the document.

To display a Gantt Chart Graphic in a Web page

1. View the Gantt Chart and on the Standard Toolbar, click Copy Picture button to open the Copy Picture Dialog box.

2. Select to GIF Image File option. The image field will be given the same name as the project file with a .GIF extension. (Graphic Interchange Format is a compressed graphics file format)

3. In the Timescale section, select the dates range or select As Shown on Screen to capture the timescale that is currently displayed on the screen.

4. Click OK.

5. On Menu, File, Save as HTML and select the filename and location and click Save.

6. In the Export Mapping Dialog box, select the map file that you want to modify and click on Edit.

7. In the HTML options section, check the option Include Image File in HTML Page and Project will display the information for the Gantt Chart GIF image.

8. Click OK and from the Export Mapping Dialog box, click Save.

9. Open the saved document in the browser to view it.

Hyperlinks

As we described earlier, hyperlink fields navigate you to another document or web site from your project. You can add hyperlinks to tasks, resources or assignments in any of the Project Views.

To add a Hyperlink

1. Open a view.

2. Select the task, resource or assignment that you want to add a hyperlink to and open Menu, Insert, Hyperlink to open the Insert Hyperlink Dialog box.

3. Enter a Web site URL or a filename from Microsoft Word or Excel in the Type the File or Web Page Name Box.

4. Enter information in the Text to Display and Screen Tip to Display when the user moves the mouse over the hyperlink indicator.

5. The Bookmark button enables you to insert hyperlinks to other Microsoft Project's files.

6. Click OK.

To delete a hyperlink select the task, resource or assignment and choose Edit, Clear, Hyperlinks.

Navigating to a Hyperlink from a Task or Resource

View, Table – Hyperlink displays the Hyperlink table. This allows you to navigate to a hyperlink associated with a task or resource in your project. Click the blue, underlined hyperlink to navigate to the document referenced by it.

Publishing Documents on the Internet and Intranet

Use the Web Publishing Wizard included in the Microsoft Office 97 and 2000 CD ROMs that will help you publish all your web pages in easy-to-follow-steps. To publish on the corporate Intranet, copy all HTML files to a location on the network server to view from your Intranet.

10.5 MS PROJECT TUTORIAL – LESSON 5

Once your Project is underway, it may be necessary to revise and change information about tasks and resources due to sudden and unforeseen changes in plans and deadlines. Tracking progress in an orderly manner will help keep your plan uptodate and in this lesson you will learn to use these tracking tools.

This lesson will also show you how to consolidate many projects into one and communicate with team members (workgroup) and exchange information such as team member assignments and task updates and other project data between various other applications.

1. Tracking Work on MS Project
2. Working with multiple projects
3. Workgroups
4. Exchanging project data with other applications.

Tracking Work on MS Project

Apart from being able to manage a project well, an effective project manager needs to ensure that the project goals are met on time and are within the budget. Anticipating the implications of a task that is slipping behind schedule, revising project plans, reassigning resources and finding ways to minimize the impact on time and costs require an orderly approach to tracking project progress.

As already mentioned in Lesson 3, it is important to save your project as a baseline and then later compare it with the actual schedules. The date, duration, work and cost fields are in the current schedule. The actual fields are where you enter actual dates and costs to show what actually happens.

There are baseline and actual fields for both tasks and assignments. The variance fields show the difference between the current value and the baseline value for each task. A positive variance means that your task will take longer to complete than originally scheduled.

Project calculates variances to the baseline for start and finish dates, duration, work and costs. You can display this variance by switching to tracking Gantt Chart View and selecting the Variance Table.

The Tracking Gantt Chart shows a baseline task bar in gray and blue or red bars for actual or scheduled tasks (dates). The variance table contains fields that show the baseline date and variances and you can add additional fields to compare actual and scheduled dates and interim plans.

After capturing and saving the Baseline, you can make subsequent changes and update it using the Save Baseline command on the menu.

Tracking information like Start and Finish dates, percentage of scheduled task duration completed, updating and rescheduling tasks will afford greater accuracy in finishing your project on time.

Entering Task Update Data

You can update most project data by using the Tracking table in the Task Sheet View. This view contains the Actual Start and Finish, Percent Complete, Actual Duration, Remaining Duration, — Actual Cost and Actual Work Fields.

To switch to the Task Sheet/Tracking Table View:

1. Choose View, More Views – Task Sheet and click OK.
2. Choose Views, Table – Tracking.

To enter data into a table:

1. Click the cell you want to enter data into, for example: Act. Start
2. Click the down arrow to select a date from the drop-down calendar.

To enter numeric values in the % completed and Actual Duration fields, use the up and down arrows on the spin box controls to change the values or type a number in the cell.

Tracking Toolbar

Description of the Tracking toolbar buttons:

Project Statistics	Opens a message box that displays current, baseline, actual and variance for Start, Finish, Duration, Work and Costs.
Update as Scheduled	Changes the actual dates of all or selected tasks to match the schedule.
Reschedule work	Reschedules the remaining duration on all or selected tasks to continue from the Status Date.
Add Progress Lines	Add progress lines to a Gantt Chart view to view whether the project is ahead or behind schedule.
% Complete	Sets the % complete on a task
Updates Tasks	Opens the update tasks dialog box.
Workgroup Toolbar	Opens the workgroup toolbar.

Using the Update tasks Dialog Box

Click the Update Tasks button from the Tracking toolbar and type a date. Click Set % or 100% complete only. Select Entire project/Selected Tasks and click OK.

To update Percent Complete

Select the tasks you want to update and click the appropriate % complete button on the toolbar.

Adding progress lines to the Gantt Chart

Project draws a line that connects in-progress tasks and tasks that should have started. Peaks pointing to the left represent work behind schedule and those pointing to the right represent work ahead of schedule.

To define progress lines

1. Choose Tools, Tracking – Progress lines to open the Progress Lines Dialog box.
2. Select the Always Display Current Progress line checkbox.
3. Click At Project Status Date or At Current Date to indicate where you want the project lines to appear.

4. Select Display Project Lines at Recurring Intervals Checkbox; then check Daily, Weekly, Monthly to specify a time that you want to display the progress lines.

5. Turn on Display Selected Progress lines and select dates.

6. Check the Line styles tab and select line type, color, and progress point shape, point color for the current and all progress lines.

7. Click OK to close the dialog box and apply the project lines.

Earned Value Analysis

Earned Value Analysis is a set of simple calculations that separate budget performance from work performance so you can manage costs and work in the midst of chaos.

Three task-related values used to calculate variances and ratios that drive performance and management are:

1. Budgeted cost of work schedules – (BCWS) or base line budget (baseline cost * baseline hours)

2. Budgeted cost of work performed – (BCWP) or actual hours budget (baseline cost * actual hours)

3. Actual cost of work performed – (ACWP) or actual cost incurred for tasks (actual cost * actual hours)

The three values are used to calculate two variances and two ratios that are:

1. Schedule Performance Index = Actual work performed/scheduled work

2. Cost Performance Index = Baseline budget costs/actual costs

These ratios that can be positive or negative numbers and are important values. The larger the number the greater the variance between performance and schedule. The Earned Value Table, applied to a task view like the Gantt Chart, lists all the relevant earned value data. For more precise definitions and help on this analysis, refer to Project's online help.

Working with Multiple Projects

Project allows you to manage smaller files individually and then combine and link subprojects when necessary. With project, you can have multiple projects open at once and switch back and forth between them with ease.

Projects come in various sizes, but regardless of the size of the overall project, it may be more convenient to work with smaller subprojects and then link them together through one master project.

To divide a project into smaller pieces

1. Open the large project that you want to subdivide.
2. Create a view that puts all the tasks you want to move in consecutive order.
3. Select these tasks and click the Cut button.
4. Click the New button and create a new project.
5. Click in the first cell of the Gantt Chart and click the Paste button.
6. Save the new project.
7. Repeat steps 3 – 6 for all the task groups that you want to move.

A master project is one that contains other projects. You can consolidate a number of projects into a master project that can be linked to subprojects in these ways:

- Consolidate multiple projects into a new project file
- Insert projects into an existing project file
- Open a resource pool and add it to a master project

You can create a master project that handles all projects within an organization and divide it into subprojects and then create reports, apply views and analyze data across each project.

To create a link between tasks that cross projects

1. Open both the projects and choose Window, Arrange All to see the two projects in the same window.
2. Double – click the task that you want to name as the successor and this will open the Task ID Dialog box.
3. Click the predecessors' tab to establish the task as the predecessor.
4. Select the ID field and type the project name and ID # of the predecessor task using this format: Project Name/Task ID #.
5. Press Enter. Project automatically adds External Project to the Type and Lag fields.
6. Click OK.

Workgroups

Communication between the project manager and team members is vital to tracking and managing any project. Regular update of a project's status within a workgroup is the ideal way to keep a project on track. A workgroup comprises people who are directly involved in the project – the manager and team in a project and external contractors and customers.

You can communicate project information electronically vie email or Intranet or Internet and set up various efficient methods of messaging between a workgroup. When using the Intranet, all workgroup members need access to a network and to the network's Web server. Team members do not require to have Project 2000 installed on their systems, only the workgroup manager must have it installed.

To configure workgroup settings

1. Select Tools, Options to open the Options Dialog box.
2. Select the Workgroup tab to choose an option from the Default Workgroup Messaging for Resources drop-down list.
3. Select Email/Web (Type the URL for the Web).
4. Identification of Web Client Server prompt appears. Select the Create Account Button and following the directions on the General tab or as given by the system administrator.
5. Click OK.

Both email and the Web can be used to send messages. On the web-based system, team members use Microsoft Project Central to read and reply to messages.

Exchanging Project Data with Other Applications

If you want to use a resource list from your Human Resources Department and a task from a workgroup leader in Project 2000, instead of manually typing in the information, you can import the data.

Importing moves data from another application into project while exporting saves project data so that it can be used in other applications. Importing data directly prevents typing errors and saves time. The copy and paste facilities of the Windows Clipboard help link files from Project to Excel or Access.

Using the Save as Menu commands, entire projects or selected data can be transferred in formats other than Project 2000's native MPP format.

Files Formats supported by Project are

- Project Database (*.mpd)
- Microsoft Project 98 (*.mpp)
- Microsoft Access 8.0 Database (*.mdb)
- Microsoft Excel Workbook (*.xls)
- Microsoft Excel Pivot table (*.xls)
- Hypertext Markup Language (*.htm)
- Text files

To import a file

1. Select the data to be imported.
2. Map the data's relationship to project data files (data exchange map).
3. Import the data.

Select the file you want to import and click open. After clicking the Open button in the Open Dialog box, the Import Mapping Dialog box is opened. In the Import File text box, is the name of the file you want to import. The Import Dialog box gives you the options of importing the entire project or importing Selective Data.

We have encountered the Define Import/Export Map Dialog box that was used to create an Export map in an HTML document in Lesson 4. It allows you to customize the Import Settings.

In addition to exporting or importing data using the File, Open or the File, Save as Command, you can also use the Edit, Copy and Edit, Paste commands to copy selected data from one document to another.

If you use the Edit, Paste Special command, you can choose the Paste Link option to create a link between the copy data you paste in the Project document and the external data.

If you select the Link check box when pasting an object into a project, the object is linked to its original application and will change if the original object is modified.

This is the end of the Microsoft Project Tutorial and you will find that you can now create reports and review your work and communicate your project successfully with others.

When you finish a project, it would help identify those things that may be usable in future projects. Keep a summary of each project you manage and use your experience to guide your decision-making about new ventures.

REVIEW QUESTIONS

1. Write short note on Project 2000 or any other SPM software.
2. Explain the main features of MS-PROJECT 2000. Where and how is it used?

APPENDICES

Appendix A

10 Best Practices for Successful Project Management

The right mix of planning, monitoring, and controlling can make the difference in completing a project on time, on budget, and with high quality results. These guidelines will help you plan the work and work the plan. Given the high rate of project failures, you might think that companies would be happy to just have their project finish with some degree of success. That's not the case. Despite the odds, organizations expect projects to be completed faster, cheaper, and better. The only way that these objectives can be met is through the use of effective project management processes and techniques. This list outlines the major phases of managing a project and discusses key steps for each one.

PLANNING

1. Plan the work by utilizing a project definition document

There is a tendency for IT infrastructure projects to shortchange the planning process, with an emphasis on jumping right in and beginning the work. This is a mistake. The time spent properly planning the project will result in reduced cost and duration and increased quality over the life of the project. The project definition is the primary deliverable from the planning process and describes all aspects of the project at a high level. Once approved by the customer and relevant stakeholders, it becomes the basis for the work to be performed. For example, in planning an Exchange migration, the project definition should include the following:

- **Project overview:** Why is the Exchange migration taking place? What are the business drivers? What are the business benefits?
- **Objectives:** What will be accomplished by the migration? What do you hope to achieve?
- **Scope:** What features of Exchange will be implemented? Which departments will be converted? What is specifically out of scope?
- **Assumptions and risks:** What events are you taking for granted (assumptions), and what events are you concerned about? Will the right hardware and infrastructure be in place? Do you have enough storage and network capacity?
- **Approach:** How will the migration project unfold and proceed?
- **Organization:** Show the significant roles on the project. Identifying the project manager is easy, but who is the sponsor? It might be the CIO for a project like this. Who is on the project team? Are any of the stakeholders represented?
- **Signature page:** Ask the sponsor and key stakeholders to approve this document, signifying that they agree on what is planned.

Initial effort, cost, and duration estimates: These should start as best-guess estimates and then be revised, if necessary, when the work plan is completed.

PROJECT WORKPLAN

2. Create a planning horizon

After the project definition has been prepared, the workplan can be created. The workplan provides the step-by-step instructions for constructing project deliverables and managing the project. You should use a prior workplan from a similar project as a model, if one exists. If not, build one the old-fashioned way by utilizing a work-breakdown structure and network diagram.

Create a detailed workplan, including assigning resources and estimating the work as far out as you feel comfortable. This is your planning horizon. Past the planning horizon, lay out the project at a higher level, reflecting the increased level of uncertainty. The planning horizon will move forward as the project progresses. High-level activities that were initially vague need to be defined in more detail as their timeframe gets closer.

PROJECT MANAGEMENT PROCEDURES

3. Define project management procedures up front

The project management procedures outline the resources that will be used to manage the project. This will include sections on how the team will manage issues, scope change, risk, quality, communication, and so on. It is important to be able to manage the project rigorously and proactively and to ensure that the project team and all stakeholders have a common understanding of how the project will be managed. If common procedures have already been established for your organization, utilize them on your project.

4. Manage the workplan and monitor the schedule and budget

Once the project has been planned sufficiently, execution of the work can begin. In theory, since you already have agreement on your project definition and since your workplan and project management procedures are in place, the only challenge is to execute your plans and processes correctly. Of course, no project ever proceeds entirely as it was estimated and planned. The challenge is having the rigor and discipline needed to apply your project management skills correctly and proactively.

- Review the workplan on a regular basis to determine how you are progressing in terms of schedule and budget. If your project is small, this may need to be weekly. For larger projects, the frequency might be every two weeks.
- Identify activities that have been completed during the previous time period and update the workplan to show they are finished. Determine whether there are any other activities that should be completed but have not been. After the workplan has been updated, determine whether the project will be completed within the original effort, cost, and duration. If not, determine the critical path and look for ways to accelerate these activities to get you back on track.
- Monitor the budget. Look at the amount of money your project has actually consumed and determine whether you're actual spending is more than originally estimated based on the work that has been completed. If so, be proactive. Either work with the team to determine

how the remaining work will be completed to hit your original budget or else raise a risk that you may exceed your allocated budget.

5. Look for warning signs

Look for signs that the project may be in trouble. These could include the following:

- A small variance in schedule or budget starts to get bigger, especially early in the project. There is a tendency to think you can make it up, but this is a warning. If the tendencies are not corrected quickly, the impact will be unrecoverable.
- You discover that activities you think have already been completed are still being worked on. For example, users whom you think have been migrated to a new platform are still not.
- You need to rely on unscheduled overtime to hit the deadlines, especially early in the project.
- Team morale starts to decline.
- Deliverable quality or service quality starts to deteriorate. For instance, users start to complain that their converted e-mail folders are not working correctly.
- Quality-control steps, testing activities, and project management time starts to be cut back from the original schedule. A big project, such as an Exchange migration, can affect everyone in your organization. Don't cut back on the activities that ensure the work is done correctly.

If these situations occur, raise visibility through risk management, and put together a plan to proactively ensure that the project stays on track. If you cannot successfully manage through the problems, raise an issue.

MANAGING SCOPE

6. Ensure that the sponsor approves scope-change requests

After the basics of managing the schedule, managing scope is the most important activity required to control a project. Many project failures are not caused by problems with estimating or team skill sets but by the project team working on major and minor deliverables that were not part of the original project definition or business requirements. Even if you have good scope-management procedures in place, there are still two major areas of scope-change management that must be understood to be successful: understanding who the customer is and scope creep.

In general, the project sponsor is the person funding the project. For infrastructure projects like an Exchange migration, the sponsor might be the CIO or CFO. Although there is usually just one sponsor, a big project can have many stakeholders, or people who are impacted by the project. Requests for scope changes will most often come from stakeholders - many of whom may be managers in their own right. One manager might want chat services for his or her area. Another might want an exception to the size limits you have placed on mailboxes. It doesn't matter how important a change is to a stakeholder, they can't make scope-change decisions, and they can't give your team the approval to make the change. In proper scope-change management, the sponsor (or a designate) must give the approval, since they are the only ones who can add funding to cover the changes and know if the project impact is acceptable.

7. Guard against scope creep

Most project managers know to invoke scope-change management procedures if they are asked to add a major new function or a major new deliverable to their project. However, sometimes the project manager doesn't recognize the small scope changes that get added over time. Scope creep is a term used to define a series of small scope changes that are made to the project without scope-change management procedures being used. With scope creep, a series of small changes - none of which appear to affect the project individually - can accumulate and have a significant overall impact on the project. Many projects fail because of scope creep, and the project manager needs to be diligent in guarding against it.

MANAGING RISK

8. Identify risks up front

When the planning work is occurring, the project team should identify all known risks. For each risk, they should also determine the probability that the risk event will occur and the potential impact on the project. Those events identified as high-risk should have specific plans put into place to mitigate them so they do not, in fact, occur. Medium risks should be evaluated to see whether they need to be proactively managed. (Low-level risks may be identified as assumptions. That is, there is potential risk involved, but you are "assuming" that the positive outcome is much more probable.) Some risks are inherent in a complex project that affects every person in the company. Other risks may include not having the right level of expertise, unfamiliarity with the technology, and problems integrating smoothly with existing products or equipment.

9. Continue to assess potential risks throughout the project

Once the project begins, periodically perform an updated risk assessment to determine whether other risks have surfaced that need to be managed.

10. Resolve issues as quickly as possible

Issues are big problems. For instance, in an Exchange migration, the Exchange servers you ordered aren't ready and configured on time. Or perhaps the Windows forest isn't set up correctly and needs to be redesigned. The project manager should manage open issues diligently to ensure that they are being resolved. If there is no urgency to resolve the issue or if the issue has been active for some time, it may not really be an issue. It may be a potential problem (risk), or it may be an action item that needs to be resolved at some later point. Real issues, by their nature, must be resolved with a sense of urgency.

APPENDIX B

CASE STUDY

CASE STUDY–1: VALLEYSPEAK PROJECT SERVER

INTRODUCTION

CaseCentral is the market leader in electronic discovery and litigation management solutions for major law firms and corporations. The company is chosen to handle many of the largest, most complex litigation projects in the nation. CaseCentral provides the most widely used online document repository, which hosts massive document collections for review by legal counsel, providing anytime, anywhere access for large, dispersed litigation review teams. The company is also rapidly emerging as a capacity leader in the electronic discovery services market with fast, affordable, high quality electronic file conversion services. Headquartered in San Francisco, CaseCentral has worked with 81 of the largest 100 US law firms (the American Lawyer Top 100) and the legal departments of hundreds of leading corporations worldwide.

THE PROBLEM

Being an Application Service Provider (ASP), scalability, reliability and redundancy are very important to Case Central when selecting the tools they work with. They have a server farm of over 200 machines scattered across multiple locations. All of these servers are managed by Case Central's IT department. Projects in IT department have very high visibility within the organization. Engineering and marketing departments rely heavily on IT projects. It is therefore natural that they are interested in tracking IT projects.

Most of the project management within the IT department in the past was done by Project Managers using Microsoft Project. They would talk to all the Team Members in a weekly status meeting and update the Microsoft Project files manually. They would then email the updated project files to the team and the company executives. This was an extremely time consuming process. In addition to requiring each person to install Microsoft Project on their machine, it required executives to manually open multiple files to see their status. Most executives therefore preferred to call the Project Managers and ask for status, and the Project Managers wasted a significant amount of time communicating the schedule status.

THE SOLUTION

It was quite obvious to IT Project Manager, Navdeep Singh, that they needed a centralized project management solution. Being a heavy user of Microsoft products their first choice was Microsoft Project Server. However after doing a pilot project it became obvious to them that Microsoft Project Server will not scale to their needs. Despite the fact CaseCentral is highly proficient in the Windows environment (since it's the primary platform for their application), just installing Microsoft Project Server was a daunting challenge. It took them longer than a week to set up Microsoft Project Server, and later it needed constant bug fixes and was highly unstable. On average it needed to be rebooted once a week. It was virus infected with a virus at one time and all of the data was lost. Over time it became increasingly obvious to Navdeep that they needed to investigate into other solutions.

Most of the solutions that they looked into were either too simple or meant for 10 person companies or too complex with costs running in hundreds of thousands of dollars. After coming across ValleySpeak Project Server, they decided to give it a try. The simplicity of the product impressed them. They were able to install it in 20 minutes - it was up and running even before people realized what was going on.

Navdeep is now a big fan of ValleySpeak Project Server. "They have a great product and great support", says Navdeep, "I personally use the ValleySpeak Project Server every day and I am totally bowled over by its stability and ease of use. Our server has been up and running for more than 6 months now. Not a single minute of downtime, while our other servers keep crashing on us all the time."

Responsive and professional tech support is also one of the reasons that Navdeep likes Valley-Speak Project Server. In his words, "ValleySpeak provides great support, period. Their tech people are really on the ball. I did actually find one bug in their product; I crafted a test case and fired off an email to support. I got a reply within 20 minutes, along the lines of "yep, that's a real bug... and it looks like it's gonna be tricky to fix". Okay, fair enough, I wasn't expecting to hear anything for a while... and then the very next day... "okay, here is a work around, that will fix your problem, we have incorporated a fix and it will show up in the next release". How good is that? I've worked for companies in the past where we couldn't even come within a mile of that level of support!"

CASE STUDY–2: AMDOCS-SOFTWARE ENGINEERING FIRM IMPROVES WORKFLOW, STREAMLINES PROJECT MANAGEMENT

INTRODUCTION

Amdocs, a global provider of customer experience systems, wanted to help its IT teams work more efficiently. Project management was particularly challenging because teams worked in multiple locations and used different software development models. To improve workflow, Amdocs is standardizing its application life-cycle management processes with Microsoft Project Server 2010. It is also integrating the solution with Microsoft Visual Studio Team System 2008 Team Foundation Server and Exchange Server 2007. IT administrators benefit from deployment that is more than 50 percent faster, and Amdocs expects to cut costs by streamlining project management. Employees have Web-based access from almost any location and work more easily with a familiar interface. Managers now have better project visibility, and the company can extend the centralized solution to include more users and applications.

SITUATION

Based in Ra'anana, Israel, Amdocs designs customer experience systems to help companies manage their business processes more efficiently. It also offers an array of consulting, technology, and product support services. Amdocs has approximately 17,500 employees, who serve its customers in more than 60 countries.

Amdocs wanted to improve the workflow for its own IT staff, which was also spread across diverse locations. For example, the team that handled application life-cycle management (ALM) was based in Israel, and the research and development and product delivery teams were located in India and the United States. In addition to working in different time zones, the IT teams used a variety of project management methodologies and tools, including Scrum for agile software development, the Waterfall development model, and Microsoft Office Project Server 2007. Also, all of the IT groups were using Microsoft Exchange Server 2007 Enterprise Edition and Microsoft Office SharePoint Server 2007 for collaboration along with Microsoft Visual Studio Team System 2008 Team Foundation Server for application life-cycle management.

- It was more like a mix-and-match than a methodology," says Tomer Simon, IT Development Manager at Amdocs. "It was hard to get an accurate status report. We had difficulty knowing exactly how much work remained, and, as a result, how close we were to meeting a deadline. Resource planning was also a big problem."

Amdocs wanted a solution that would connect its infrastructure and its project teams in multiple locations. "We were looking for a platform that could provide a single view of all of the other systems in one place, including information about a project's development status, change configuration, and resources," says Simon. "We were looking for one solution that could handle almost everything."

SOLUTION

The company's ALM team in Israel had used Microsoft Office Project Server since the product was first released. The team planned to upgrade as usual to the next version-Microsoft Project Server 2010. But when it learned more about Project Server 2010, it realized it could use the solution to standardize project management for multiple IT teams.

Because Project Server 2010 is built on Microsoft SharePoint Server 2010, Amdocs would be able to implement a unified project and portfolio management solution that includes the familiar SharePoint Server 2010 user interface, a common data store, and enhanced business intelligence and reporting tools. Equally important, Amdocs could take advantage of the comprehensive application programming interface to integrate the solution with other applications.

Amdocs successfully completed a pilot project with Project Server 2010 and began deploying the solution in its production environment in December 2009. Orantech Technologies, a Microsoft Gold Certified Partner, used its Project Party software to connect Visual Studio Team System 2008 Team Foundation Server with Project Server 2010. Orantech is also helping with ongoing implementation and training, and Amdocs expected to roll the solution out enterprisewide by April 2010.

In addition to Team Foundation Server, Amdocs integrated Project Server 2010 with Exchange Server 2007 so that employees would be able to update tasks in the Microsoft Office Outlook 2007 messaging and collaboration client. It is also implementing Microsoft Business Intelligence tools such as Excel Services in Microsoft SharePoint Server 2010. Simon says, "Because Project Server 2010 is based on SharePoint Server 2010 and we've integrated it with Team Foundation Server, we can play around with different Web Parts and create a site that shows us the status from multiple systems on one platform."

Assaf Leibovitch, Senior Developer at Amdocs, adds, "From a developer's point of view, deploying Project Server 2010 was very intuitive. The documentation from Microsoft was also specific and detailed, and if we follow the instructions, it just flows."

All Windows-based and Web-based client applications for Microsoft Project 2010 use the Project Server Interface, which is a set of Web services built on the Microsoft .NET Framework 3.5 and Windows Communication Foundation. Amdocs used the interface to integrate the customized applications it had designed for Office Project Server 2007. The company reports that the conversion went smoothly, with little effort required from developers.

ALM program managers in Israel typically work with the Microsoft Project Professional 2010 client. Amdocs is also using the new Project Server Interface to extend the solution of more users. For example, team leaders in India and other locations can use Microsoft Project Web App to connect through the Internet and edit projects. In addition, Amdocs uses Windows Communication Foundation to configure Web services that work with multiple communication protocols and connect with more applications.

BENEFITS

With Project Server 2010, Amdocs can ease deployment and cut costs by simplifying administration. It has centralized project management, and connected more applications and employees.

Speeds Deployment by More Than 50 Percent

In the past, it could take Amdocs several days to upgrade Project Server and ensure that earlier projects would work with the new version. However, Amdocs can now deploy Project Server 2010 in less than a day, "It's at least 50 percent faster to deploy Project Server 2010 compared with earlier versions," says Simon. "We also needed an external vendor to help us before, but now we can do it ourselves".

Working with old projects from earlier versions is also easy. "If we want to open a file from an older version in Project Server 2010, we just open it, save it, and that's it—it works immediately," says Simon. "In other words, there are no migration problem, no conversion issues, and no errors."

Reduces Total Cost of Ownership

Although Amdocs is still in the early stage of deployment, it already expects to cut costs buy streamlining management. Before, Amdocs needed a dedicated administrator for Office Project Server 2007. Now, the administrator can focus on other projects. "Because Project Server 2010 is built on SharePoint Server 2010, we only need one person to manage the shared infrastructure," says Simon.

- The improvements with connectivity and the backward compatibility also make development faster and easier, because we have more options available," he continues. "So it takes approximately 25 percent less time to design connectors or software that works with Project Server 2010, and the bottom line is that the process then becomes less costly."

Improves Workflow with Familiar Interface

By connecting Project Server 2010 with other applications such as Team Foundation Server, Amdocs IT development personnel can work more efficiently and gain better control of projects. "As a manager, the biggest benefits of Project Server 2010 are the improved visibility and the ability to use one tool to manage the entire process," says Simon. "I can quickly see the status or cost of a project, without needing to extract information and manipulate it in another program."

The solution connects with applications that Amdocs relies on, and it also integrates more easily into employees' workflow. As a result, they are more likely to take advantage of its capabilities.

"With its familiar interface, Project Server 2010 is easy to work with, even for a new user," says Simon. "It's now much more intuitive and straightforward to just start planning."

Increases Scalability

Amdocs is using the new Project Server Interface to design a unified project and portfolio management solution that performs better and takes advantage of more resources. Simon says, "We have a lot of servers out there, and we discovered that we can connect and extract data more easily with Project Server 2010 without worrying about protocol conversion or performance issues."

The company can also extend the solution and benefits of a connected infrastructure to more people. For example, Amdocs can use built-in features and technologies such as Microsoft Project Web App to facilitate Internet-based access for a growing number of employees. As a result, the company can increase scalability easier while taking advantage of its existing infrastructure.

Amdocs also has the flexibility it needs to integrate Project Server 2010 with other applications. As a result, team leaders working offsite or in remote locations can work with the same data repository and many of the same management tools as local employees. Leibovitch says, "With Windows Communication Foundation support in Project Server 2010, we can scale up and out more easily."

CASE STUDY–3: CRONUS CONSULTANCY FOR SOFTWARE DEVELOPMENT

This case study is presented by Delio Nobrega, who works as a business consultant for Cronus Consultancy Limited - a software development company based in the Channel Islands.

Company Profile

Cronus is a Microsoft Software Development Gold Partner. We write web based, windows and server based systems for local government and finance clients. We have also written products which we sell. For example, a Trust Administration System called OSCAR (www.oscar-e.com).

We offer a variety of services, such as Business Analysis, Consultancy, Database Administration, Knowledge Management, Web Development and others. Full list of our services can be found at www.cronus.biz

Challenge

Before we started using Easy Projects .NET we had been managing our projects manually using a combination of Outlook, Excel and paper. With about 10 projects going on simultaneously, the biggest challenge was keeping track of work undertaken on bugs that are currently active, assignment of work and allowing the client a look at what stage their project was at.

Solution and Results

After evaluating 3 other project management tools, we chose Easy Projects .NET. As the name implies it was very straightforward to implement, and we wanted to purchase a product that came with a source code option also.

Right now 11 out of 12 people in our company are using EP.NET. The most important features of EP.NET for us are its ability to assign tasks to people and its ability to add project documentation.

The main benefits for Cronus are that it has enabled us to give our clients the ability to see progress against the issues or requests they log. This has saved us time and it also has made us more structured.

Internally it has allowed the project managers to easily allocate and track resourcing across their own project, while still keeping an active role in the project. It has helped immensely and has made us think of new enhancements that could be implemented. That leads me onto the final benefit of using EP.NET, the development and support staff have been excellent and are always willing to take on board new features and very helpful.

CASE STUDY–4: INTERACTIVE PALETTE, INC. FOR WEB DESIGNING AND DEVELOPMENT

This case study is presented by Kevin McNally - the founder of the USA-based Interactive Palette, Inc.

Company Profile

Interactive Palette is an independent consulting firm based in the greater Boston area offering web design and development with a specialization in Flash Product and Promotional Presentations and Search Engine Optimization. We have proven strengths in delivering award-winning solutions to a diverse client base.

If you're small to mid-sized company wants to enhance an existing web site, establish an internet presence or launch a website marketing campaign, Interactive Palette will work within your needs and constraints to create custom, compelling website designs that deliver the digital solutions you demand. Full list of our services can be found at www.interactivepalette.com

Challenge

Before we acquired Easy Projects .NET, we were not keeping track of time properly and a lot time wasn't being billed.

Solution and Results

Have evaluated 4 different project management applications we decided to purchase Easy Projects .NET mainly because it had some of the features we liked from MS Project and I really like the possibility of customizing the product with some of our other forms.

Currently 5 employees are actively using EP.NET managing 15+ projects simultaneously. We find time tracking, reporting and the organization of projects and tasks to be most important features for us. Right now we are about to start using the bug tracking feature in Easy Projects for QA purposes on websites, which we hadn't planned on while choosing a pm tool.

The key benefit of using Easy Projects .NET in our company - now when we want to know how profitable a project was we run a report and when it is time to do billing it is all right there. This was a big problem to us before and billing to take forever.

APPENDIX C

SUGGESTED EXPERIMENTS IN SPM LAB

Experiment No. 1

Suppose that a company wants to build software for railway reservation system. Write the scope of statement that describes the software (RRS).

Experiment No. 2

Estimate the effort and cost required to build the above software. Use any estimation technique.

Experiment No. 3

Develop SRS for railway reservation system.

Experiment No. 4

Draw ER diagrams, DFDs till level-2 for Railway Reservation System (RRS).

Experiment No. 5

Design a data dictionary for RRS.

Experiment No. 6

Develope a user interface design for RRS.

Experiment No. 7

Compare the project life cycle of the following:

(*a*) An engineering product, say a nuclear power station project.

(*b*) An housing project.

(*c*) A computer installation project.

Experiment No. 8

Prepare a project life cycle chart.

*A*PPENDIX D

AREAS UNDER THE NORMAL CURVE

	00	.01	.02	.03	.04	.05	.06	.07	.08	.09
0.0	.50000	.50399	.5079*8	.51197	.51595	.51994	.52392	.52790	.53188	.53586
0.1	.53983	.54380	.54776	.55172	.55567	.55962	.56356	.56749	.57142	.57535
0.2	.57926	.58317	.58706	.59095	.59483	.59871	.60157	.60642	.61026	.61409
0.3	.61791	.62172	.62552	.62930	.63307	.63083	.64058	.64431	.64803	.65173
0.4	.65542	.65910	.66276	.66640	.67003	.67364	.67724	.68082	.68439	.68793
0,5	.69146	.69497	.69847	.70194	.70540	.70884	.71226	.71566	.71904	.72240
0.6	.72575	.72907	.73237	.73536	.73891	.74215	.71537	.74857	.75175.	.75490
0.7	.75801	.76115	.76424	.76730	.77035	.77337	.77037	.77935	.78230	.78524
0.8	,78814	.79103	.79389	.79673	.79955	.80234	.80511	.80785	.81057	.81327
0.9	.81594	.81859	.82121	.81381	.82639	.82894	.83147	.83398	.83646	.83891
1.0	.84134	.84375	.84614	.84849	.85083	.85314	.85543	.85769	.85993	.86214
1.1	.86433	.86650	.86864	.87076	.87286	.87493	.87698	87900	.88100	.88298
1.2	.88493	.88686	.88877	.89065	.89251	.89435	.89617	.89796	.89973	.90147
1.3	.90320	.90490	.90658	.90824	.90988	.91149	.91309	.91466	.91621	.91774
1.4	.91924	.92073	.92220	.92364	.91S07	.92647	.92785	.92922	.93056	.93189
1.5	.93319	.93448	.93574	.03699	.93822	.93943	.94062	.94179	.94295	.94408
1.6	.94520	.94630	.94738	.94845	.94950	.95053	.95154	.95254	.95352	.95449
1.7	.95543	.95637	.95728	.95818	.95907	.95994	.96080	.96164	.96246	.96327
1.8	.96407	.96485	.96562	.96638	.96712	.96784	.96856	.96926	.96995	,97062
1.9	.97128	.97193	.97157	.97230	.97381	.97441	.97500	.97558	.97615	.97670
2.0	.97715	.97784	.97831	.97882	.97932	.97982	.08030	.98077	.98124	.98169
2.1	.98214	.98157	.96300	.98341	.98382	.98122	.98461	.98500	.98537	.98574
2.2	.98610	.98645	.98679	.98713	.98745	.98778	.98809	.98840	.98870	.98899
2.3	.98928	.98956	.98983	.99010	.99036	.99061	.09086	.99111	.99134	.99158
2.4	.99180	.99202	.99224	.99245	.99266	.99286	.99305	.99324	.99343	,99361
2.5	.99379	.99396	.99413	.99430	,99446	.99461	.99477	.99492	.99606	.99520
2.6	.99534	.98547	.99560	.99573	.99585	.99598	.99609	.99621	.99632	.99643
2.7	.99653	.99664	.99674	.99683	.99693	.99702	.99711	.99728	.99728	.99738
2.8	.99744	.99752	.99760	.99767	99774	.99781	.99788	.99795	.99801	.99807
2.9	.99813	.99819	.99815	.90631	.99836	.99841	.99846	.99851	.99856	.99864
3.0	.09865	.99869	.99869	.99874	.99882	.99886	.99899	.99893	.99826	.99900
3.1	.99903	.99906	.99910	.99913	.99916	.99918	.99921	.99924	.99926	.99929
3.2	.99931	.99934	.99936	.99938	.99940	.99942	.99944	.99946	.99948	.99950
3.3	.99952	.99953	.99955	.99957	.99958	.09960	.99961	.99962	.99964	.99965
3.4	.99966	.99968	.99969	.99970	.99971	.99972	.99973	.99974	.99975	.99976
3.5	.99977	.99978	.99978	.99979	.99980	.99981	.99981	.99982	.99983	.99983
3.6	.99984	.99985	.99985	.99986	.99986	.99987	.99967	.99988	.99988	.99989
3.7	.99989"	.99990	.99990	.99990	.99991	.99991	.99992	.99992	.99992	.99992
3.8	.99993	.99993	.99993	.99994	.99994	.99994	.99994	.99995	.99995	.99905
3.9	.99995	.99995	.99996	.99996	.99996	.99996	.99996	.99996	.99997	.99997

Appendix E

Software Project Management Acronyms and Glossary

COMMON PROJECT MANAGEMENT ACRONYMS

AC	:	Actual cost
ACWP	:	Actual Cost of Work Performed
AD	:	Activity Description
ADM	:	Arrow Diagramming Method
AF	:	Actual Finish date
AOA	:	Activity-on-Arrow
AON	:	Activity-on-Node
AS	:	Actual Start date
BAG	:	Budget at Completion
BCWP	:	Budgeted Cost of Work Performed
BCWS	:	Budgeted Cost of Work Scheduled
CAP	:	Control Account Plan (previously called Cost-Account Plan)
CCB	:	Change Control Board
CPFF	:	Cost-Plus-Fixed-Fee
CPI	:	Cost Performance Index
CPIF	:	Cost-Plus-Incentive-Fee
CPM	:	Critical Path Method
CV	:	Cost Variance
DD	:	Data Date
DU	:	Duration
EAC	:	Estimate to Complete
EV	:	Earned Value
EVM	:	Earned Value Management
FF	:	Free Float or Finish-to-Finish
FFP	:	Firm Fixed-Price
FRF	:	Fixed Price-Incentive-Fee
FS	:	Finish to-Start

GERT	:	Graphical Evaluation and Review Technique
IFB	:	Invitation for Bid
LF	:	Late Finish date
LOE	:	Level of Effort
LS	:	Late Start date
OBS	:	Organization (al) Breakdown Structure
PC	:	Percent Complete
PDM	:	Precedence Diagramming Method
PERT	:	Program Evaluation and Review Technique
PF	:	Planned Finish date
PM	:	Project Management or Project Manager
PMBOC	:	Project Management Body of Knowledge
PMP	:	Project Management Professional
PS	:	Planned Start date
PV	:	Planned Value
QA	:	Quality Assurance
QC	:	Quality Control
RAM	:	Responsibility Assignment Matrix
RDU	:	Remaining Duration
RFP	:	Request for Proposal
RFQ	:	Request for Quotation
SF	:	Scheduled Finish date or Start-to-Finish
SOW	:	Statement of Work
SPI	:	Schedule Performance Index
SS	:	Scheduled Start date or Start to Start
SV	:	Scheduled Variance
TC	:	Target Completion date
TF	:	Total Float or Target Finish date
TQM	:	Total Quality Management
TS	:	Target Start date
VE	:	Value Engineering
WBS	:	Work Breakdown Structure

SOFTWARE PROJECT MANAGEMENT GLOSSARY

1. **Accrual Rules.** One or more sets of instructions which determine how much, and how fast, employees can accrue vacation, sick leave, and other leave.

Activity. An element of work performed during the course of a project. An activity normally has an expected duration, an expected cost and expected resource requirement. Activities can be sub-divided into tasks.

Activity Definition. Identifying the specific activities that must be performed to produce the various project deliverables.

Activity Description (AD). A short phrase or label used in a project network diagram. The activity description normally describes the scope of work of the activity.

Actual Client Cost. Actual client cost is the calculated amount used for billing clients (or charge back) for tasks or projects.

Actual Company Cost. Actual company cost is the calculated amount used to determine your company's cost to accomplish a task or project.

Actual Finish Date (AF). The project in the time that work actually ended on an activity.

Actual Start Date (AS). The point is time that work actually started on an activity.

Application Area. A category of projects that have common elements not present in all projects. Application areas are usually defined in term of either the product of the project (*i.e.,* by similar technologies or industry sectors) or the type of customer.

Assumptions. Assumptions are factors that for planning purposes, are considered to be true, real or certain. Assumptions affect all aspects of project planning, and are part of the progressive elaboration of the project.

Audit Trail. A log of additions, changes or deletions to numerous types of records in a database.

Bar Chart. A graphic display of schedule-related information. In the typical bar chart, activities or other project elements are listed down the left side of the chart, dates are shown across the top, and activity durations are shown as date-placed horizontal bars. Also called a Ghantt chart.

Change Control Board (CCB). A formally constituted group of stakeholders responsible for approving or rejecting changes to the project baselines.

Checklists. A listing of many possible risks that might occur on a project. It is used as a tool in the risk identification process. Checklists are comprehensive, listing several types of risk that have been encountered on prior projects.

Component. A constituent part, an element.

Constraint. Applicable restriction that will affect the performance of the project. Any factor that affects when an activity can be scheduled.

Control Account Plan (CAP). CAP is a management control point where the integration of scope and budget and schedule takes place, and where the measurement of performance will happen. CAP's are placed at selected management points of work breakdown structure.

Cost Budgeting. Allocating the cost estimates to individual work activities.

Cost Control. Controlling changes to the project budge.

Data Date (DD). The data at which, or up to which, the project's reporting system has provided actual status and accomplishments. Also called as-of-date.

Decision Tree Analysis. The decision tree is a diagram that describes a decision under consideration and the implications of choosing one or another of the available alternatives. It incorporates probabilities or risk and the costs on rewards of each logical path of events and future decisions.

Early Finish Date (EF). In the critical path method, the earliest possible point in time on which the uncompleted portions of an activity (or the project) can finish, based on the network logic and any schedule constraint. Early finish dates can change as the project progresses and changes are made to project plan.

Early Start date (ES). In the critical path method, the earliest possible point in time on which the uncompleted portions of an activity (or the project) can start, based on the network logic and any schedule constraints. Early start dates can change as the project plan.

Effort. The no. of labor units required to complete an activity, or other project elements. Usually expressed as staff hours, staff days, or staff weeks.

ETC. (Estimate to complete) is optional data fields used by employees to enter how many hours they estimate are required to complete the task.

Fast Tracking. Compressing the Project schedule by overlapping activities that would normally be done in sequence, such as design and construction.

Float. The amount of time than an activity may be delayed from its early start without delaying the project finish date. Float is a mathematical calculation, and can change as the project progresses and changes are made to the project plan. Also called slack, total float, and path float.

Free Float (FF). The amount of time than an activity can be delayed without delaying the early start of any immediately following activities.

Functional organization. An organization structure in which staff are grouped hierarchically by specialty (*e.g.* production, marketing, engineering, and accounting at the top level; with engineering further divided.)

Grade. A category or rank used to distinguish items that have the same functional use (*e.g.* "hammer"); but do not share the same requirements for quality (*e.g.*, different hammers may need to withstand different amounts of force).

Items list. A tab within the Project Workspace view which displays clients, Projects, Employees etc. in a grid type view.

Job Scheduler. Job scheduler is a job automation tool. It gives the user the ability to automate functions such as sending reports and notices, or maintaining the database.

Life-cycle Costing. The concept of including acquisition, operating, and disposal costs when evaluating various alternatives.

Limited Access. A subset of security options applied to certain views which limit the views to only those items (projects, tasks, task groups, teams, etc.) for which they are the designated 'manager'.

Logical Relationship. A dependency between 2 project activities, or between a project activity and a milestone.

Network Logic. The collection of activity dependencies that makes up a project network.

Organizational Planning. Identifying, documenting and assigning project roles, responsibilities and reporting relationships.

Path. A set of sequentially connected activities in a project network diagram.

Performance Measurement Baseline. An approved plan against which deviations are compared for management control.

Performing Organization. The enterprise whose employees are most directly involved in doing the work of the project.

PERT Chart. The term is commonly used to refer to a project network diagram see program evaluation and review technique for the traditional definition of PERT.

Product Scope. The features and functions that characterize a product or service.

Project Approvals. Project Approval is a mechanism allowing a manager or system administrator to create approval workflow's that are project specific.

Project Life Cycle. A collection of generally sequential project name whose name and no. are determined by the control needs to the organization or organizations involved in the project.

Project Management Body of Knowledge. An inclusivc term that describes the sum of knowledge within the profession of project management. As with other professions such as law, medicine, and accounting– the body of knowledge rests with the practitioners and academics that apply and advance it. The PMBOK includes proven traditional practices that are widely applied, as well as innovative and advanced ones that have seen more limited use.

Project Management Professional (PMP®). An individual certified as such by the Project Management Institute (PM I®)

Project Phase. A collection of logically related project activities, usually culminating in the completion of a major deliverable.

Qualitative Risk Analysis. Performing a qualitative analysis of risks and conditions to prioritize their effects on project objectives. It involves assesst the probability and impact of project risk(s) and using methods such as the probability and impact matrix to classify risks into categories of high, moderate, and low for prioritized risk response planning.

Risk. An uncertain event on conditions that, if it occurs, has +ve or –ve effect on a projects objectives.

Risk Management Planning. Deciding how to approach and plan, risk management activities for a project.

Schedule Development. Analyzing activity sequences, activity durations and resources requirements to create the project schedule.

Scope. The sum of the products and services to be provided as a project.

Staff Acquisition. Getting needed human resources assigned to and working on the project.

Statement of Work (SOW). A narrative description of products or services to be supplied under contract.

Tasks. Tasks are activities or work efforts against which employees track their time. Tasks are fundamental components required for time tracking.

Total Quality Management (TQM). A common approach to implementing a quality improvement program within an organization.

Value Engineering (VE). It is a creative approach used to optimize life cycle costs, save time, increase profits, improve quality, expand market share, solve problems, and for use resources more effectively.

WBS (Work Breakdown Structure). A deliverable-oriented grouping of project elements that organizes and defines the total work scope of the project.

Appendix F

Small Projects

E.1 INTRODUCTION

Many of the readers of this book are students who will have to plan their own projects that they will have to carry out as part of their course of studies. In some cases, these will be undertaken for an external client. In other cases, a piece of software, perhaps of an experimental nature, is to be produced where there is no identifiable client, apart from a project tutor. Although the projects that are carried out for 'real' clients are more convincing tests of the student, they are in many ways more risky than the purely academic ones.

One of the problems that students face when planning projects is applying techniques that they have learnt on software project planning courses and that were designed for much larger-scale projects than their own. In this chapter we present an outline of how students should set out their plans. It is based on a structure that we have recommended to our own students over the years. The overall Step Wise approach is still applicable to the planning of your project – but the techniques used at the different steps of the planning process will need to be carefully chosen as appropriate to the scaled-down application.

E.2 SOME PROBLEMS WITH STUDENT PROJECTS

There are some problems or risks that seem particularly to affect student projects.

Use of unfamiliar tools

Very often students will be using a new software tool (for example, an application builder in a Windows environment) that they have not used before. Clearly, time will need to be allocated in the project plan to learning the package. When trying to formulate plans, because of their ignorance of the software tool, students might have difficulties estimating how long tasks will take. There will also be risks that unexpected technical problems will halt or delay the project's progress. Students often have other things on their minds, such as examinations, in the period leading up to the start of the project, but the risks of technical problems will be reduced if they are able to try out the software tools at this point.

If you are really keen to learn about and use a new tool, you might consider framing the objectives of the project to include an evaluation of the tool. Technical difficulties then become transformed from being obstacles to being interesting data.

Uncertain design requirements

Many project assignments require students to demonstrate a careful analysis and design process and then to build at least part of the application. Until the analysis has been done, it can be difficult to plan exactly how design and software building is to be executed.

Two points are worth making here. The first is that the structure of the project is going to be dictated to a large extent by the amount of time available.

Say that you have ten weeks in which to carry out a project that involves analysing user requirements, designing a new system and building it. It is not a bad idea to start planning on the basis of doing the project in nine weeks in order to allow for slippage. Working backwards from the end, of those nine weeks, it might be decided that the last two should be reserved for testing and evaluation. This may seem an excessive amount of time to some students, but with most assessment schemes your project is likely to gain really good marks if you can demonstrate that it is of good quality and that you have evaluated it carefully. You might decide to allocate proportions of the remaining first seven weeks to analysis, design and system building so that you get the following skeleton schedule:

- **weeks 1–2:** analysis;
- **week 3:** design;
- **weeks 4–7:** system building;
- **weeks 8–9:** testing and evaluation;
- **week 10:** contingency.

The actual breakdown will vary, depending on the circumstances of your particular project. A very well structured problem area means that you can spend less time on analysis, while, on the other hand, if you know the software building tools that you are going to use, you might decide to reduce the time for analysis in favour of devoting more effort to a more sophisticated operator interface.

The second point, and one that we have already emphasized, is that you should be prepared to delay planning a particular phase of a project in detail until more information becomes available. When you have completed your analysis phase at the end of the second week in the above plan, you should be in a much better position to plan the design and system building phases in more detail – you will know, for example, what the main transactions are going to be in an information systems application. You should also be in a position to cut down the scope of the application to fit the time available to build it!

Incomplete systems

Sometimes students simply run out of time and so do not have a working system to demonstrate. In the case of student projects, it is a good idea to try to arrange things so that you have something, even if it is not much, to demonstrate from a relatively early stage of the project. In the skeleton plan described above, it would not be a bad idea to have certain features of the application up and running after the first week or so of system building. Having these 'in the bag' so to speak, you can go on and add new functions or features, secure in the knowledge that at any point you will have at least something to demonstrate. You might find that you can even break the work down into increments to give you something similar to the following:

- **weeks 1-2:** analysis;
- **week 3:** design increment 1;
- **week 4:** build/test increment 1;
- **week 5:** design increment 2;

- **week 6:** build/test increment 2;
- **week 7:** design increment 3;
- **week 8:** build/test increment 3;
- **week 9:** evaluate complete system;
- **week 10:** writing up/contingency.

If something went wrong so that increment 3 could not be completed on time, you would still have increments 1 and 2 to demonstrate and in your project evaluation report you could describe your proposed design for increment 3.

In the skeleton plan above, we have allocated week 10 for writing up. As a general rule, it is best to try and write up as much of your project as you can as you go along. It will not save you any time if you leave it and you will be able to write more clearly and fully about your analysis process, for example, while it is still clear in your mind. Pausing to do this writing should also help you to reflect on what you are doing and help you consider coolly what needs to be done next.

Lack of commitment from clients

In most cases, where a project is for an external client, the student will not be being paid. The advantage of this is that the outside organization might be attracted to the chance of having a free resource and take on a project for a student about whose capabilities they know little or nothing. The danger is that because the resource is free, there will be little commitment from the client to the project. If they had to pay for the work, they would think more carefully about whether they really needed the work before agreeing to the project and they would pay attention to getting value for money.

There is no such thing as a completely free project, even when the student's efforts are free. The student might need access to hardware and software facilities to be provided by the client and even where this is not the case, the client needs to be prepared to give up some spare time to discuss requirements and to evaluate intermediate results. It has to be said that sometimes clients can let students down as far as this is concerned. Our advice to students would be to try and get to know your client thoroughly before the project starts. Try, for instance, to meet all the people in the organization who will be affected by your work before you actually get going on the project. Show a copy of your initial project plan to the client and get them to comment on it. Arrange beforehand to have regular meetings with them to discuss progress.

E.3 CONTENT OF A PROJECT PLAN

We discuss below what should be in your project planning document. It is suggested that the plans for small projects being done for outside clients should follow this format. Specific course requirements (for example, learning objectives) are not covered.

1. Introduction. In the introduction to any document, it is always a good idea to explain briefly what the document is and why it has been produced. The introduction to the project plan should also include:

- identity of client – that is, the organization or department for whom the work is to be done:
- short description of project – not more than two or three brief lines:
- identity of the project authority – the person or persons within the client organization who will have authority over the project's direction. It is essential that such an authority be identified.

2. Background. This includes:

- relevant information about the client's business;
- descriptions of the existing software/hardware environment;
- circumstances or problems leading to the current project;
- work already carried out in the area of the projcct;
- stakeholders in the projects, in other words, all those who will be affected by the project or who have some other interest in it.

3. Objectives of the project. These might already have been defined in a terms of reference (TOR) document. If so, then it can be attached as an appendix to the plan.

The objectives must define what is to be achieved and the method of measuring the extent of that achievement. Our problem with small projects conducted by students is that the project's success is evaluated in course terms soon after its completion, whereas the project's true value to the users will take much longer to emerge (if ever!). As this document is in part for the benefit of the client, the objectives from the client's viewpoint should have the emphasis here.

Students sometimes have problems distinguishing between their personal objectives and those of the project. The project objectives are those that are held in common between the client and the developer in relation to the project. For example, a university student could be producing a timetabling system for a local school. The creation of this timetabling system would be a project objective shared by the developer (the student) and the client (the school). The student will also have a personal objective of getting a good grade that is not shared by the school (although they will usually be sympathetic to it). From the viewpoint of the student's personal objectives the creation of the timetabling system is seen as a sub-objective, or goal, a step on the way to achieving the overarching objective of getting the good grade.

Where there are several objectives, an order of priority should be given to them if this is possible.

4. Constraints. It is convenient to merge this into the project objectives, above. Constraints include:

- externally imposed time scales;
- legal requirements;
- specific standards;
- limitations on the people who can be approached for information.

5. Methods/technology to be used. These might have already have been laid down for you, or it might be that part of the project will involve the selection of the most appropriate technologies and methods. In other cases, you need to specify the general approach you will take – for instance, that you are going to use a structured systems analysis and design method (like SSADM), or a soft system methodology (like SSM) approach. In a small project it is unlikely that you will have time for a full structured analysis/design approach and you might decide to use a subset of the techniques – this should be specified.

The methods selected will, of course, govern much of what will go into Section 6 below on project products and Section 7 on activities. When discussing prototypes, we emphasized that a prototype should always be a tool for learning or clarifying something (for instance, the best interface for the user). If you claim you are producing a prototype, then you must be prepared to define learning objectives for the prototype, a method of evaluation and an analysis of what has been learnt.

If you are developing software, then the choice of software tools should be stated in this section.

Some justification of the decisions made here should be given, as they can be crucial to the success of the project.

The training and other resource requirements (perhaps the need to purchase a particular package) that result from the decisions should be noted.

6. Project products. This is a list of all the products or deliverables that the project will produce, such as software modules, documentation, user guides and reports. Intermediate products, such as design documents, should be included.

7. Activities. This is a list of the main activities that the project will involve. There must be activities to produce all the products listed in 6. Also, in general, each activity should result in some deliverable: avoid tasks like 'familiarization with departmental procedures' in favour of 'documentation of departmental procedures'. In identifying activities, new interim products will often be discovered. For each activity, define:

- **pre-requisites,** what has to be done before this activity can start;
- **dependent activities,** activities that need this one to be completed first;
- **estimated time/effort,** this may be a range of values;
- **quality checks,** details of how you are going to verify and validate the product of the activity.

PERT or Gantt charts may be used but are often not needed.

8. Resources. This includes staff time, accommodation and hardware/software requirements.

9. Risk analysis. Identify the main things that can be seen as possibly going wrong. Typically this might include:

- unavailability of resources (as with a delay in getting a software package);
- unavailability of key client personnel;
- technical problems (such as software bugs).

A priority can be given to each risk by allocating a *probability rating* (1–10) and a seriousness of *impact rating* (1–10). Multiplying the two together gives an overall score for priority purposes.

For the most serious risks (those with the highest scores), *preventive* measures to reduce or remove the risk should be specified. For example, in order to prevent problems with the unavailability of key client personnel, meetings must be arranged with them at the planning stage and holiday plans may be ascertained. In some cases *contingency* measures that can be undertaken once the risk has actually materialized are more appropriate.

APPENDIX G

MISCELLANEOUS NUMERICAL PROBLEMS ON SOFTWARE PROJECT MANAGEMENT

Example 1. *Compute the FP value for a project with the following information domain characteristics.*

No. of user inputs = 32
No. of user outputs = 60
No. of user inquires = 24
No. of user files = 08
No. of external interfaces = 2

Assume that all complexity adjustment values are average. Assume 14 algorithms have been counted.

Solution.

Measurement parameter	*Count*	*Weighting factor*
No. of UI	32 × 4 =	128
No. of O/ps	60 × 5 =	300
No. of 1/ps	24 × 4 =	96
No. of files	8 × 10 =	80
No. of EI	2 × 7 =	14
	Count–Total	618

$$\text{FP} = 618 \times (0.65 + 0.01 \times 14 \times 3)$$

$$\text{Function point} = 618 \times (0.65 + 0.42)$$

$$= 618 \times 1.07 = 661.26 = 661$$

and feature point $= (32 \times 4 + 60 \times 5 + 24 \times 4 \mp 80 + 14) * 1.07 + \{12 \times 15 * 1.07\} = 853.8.$

Example 2. *Suppose we are faced with developing a system that we expect 1,00,000 lines of code compute the effort and development time for each of the 3 development modes i.e. organic, semi-detached and embedded.*

Solution. Given

Size of software = 1,00,000 LOC = 100 KLOC.

(*a*) Organic mode : $E = a$ (size of $p - m$) project^b

$= 2.4\,(100)^{1.05}$

$$\boxed{E = 302.14\, p - m}$$

and $D = C\,(\text{effort})^d = 2.5\,(302.14)^{0.38} = 21.9$ months

and No. of persons $= \dfrac{302.14}{21.9}$

$= 14$ persons.

Similarly, we can compute for other two modes also.

Example 3. Suppose that you are developing a software product in organic mode. You have estimated size of product to be about 1,00,000 LOC. Compute nominal effort and development time.

Solution. Size = 1,00,000 LOC = 100 KD LOC

For organic mode,

$$a = 3.2 \text{ and } b = 1.05$$

$\therefore$ Nominal effort, $E_1 = a * (\text{size})^b$

$= 3.2 * (100)^{1.05}$

$= 3.2 * 125.8925 = 402.8561318$ p.m.

Minimum development time,

$$M = 2.5\, E^{0.38}$$

$$M = 2.5 * (402.9)^{0.38} \text{ months}$$

$$= 2.5 * 9.77 = 24.43 \text{ months.}$$

Example 4. *Prove that the manpower utilization, m (t), is given by the formula:*

$$m(t) = \frac{dy}{dt} = 2K\,at\,e^{-at^2}$$

where K is area under curve $[0, \infty]$

a is parameter affecting shape of curve.

t is elapsed time.

$\dfrac{dy}{dt}$ *is manpower utilization/time.*

Solution. $$m(t) = \frac{dy}{dt} = 2K\,at\,e^{-at^2} \qquad ...(1)$$

$\therefore$ $$\int_0^{\infty} \frac{dy}{dt} = y\,(t) = K[1 - e^{-at^2}]$$

$\therefore$ $y(t) = 0$ (at start of project)

$y(\infty) = K$ (grows monotonically)

To find relationship between 't_d' (peak time) and 'a', we find 2nd derivative *i.e.*,

$$\frac{d^2y}{dt^2} = 2\,Kac^{-at\,2}\,[1 - 2at^2] = 0$$

or $$t_d^2 = \frac{1}{2a} \quad \text{or} \quad a = \frac{1}{2t_d^2} \qquad ...(2)$$

$$m(t) = \frac{2K}{2t_d^2} te^{\frac{-t^2}{2t_d^2}} = \frac{K}{t_d^2} te^{\frac{-t^2}{2t_d^2}} \qquad ...(3)$$

Now, at $t = t_d$, peak manning is obtained

i.e., $$m(t_d) = m_0 \text{ (say)}$$

Putting $t = t_d$ in (3), we get

$$m(t_d) = \frac{K}{t_d^2} t_d e^{\frac{-t^2}{2t_d^2}} = \frac{K}{t_d} e^{\frac{-1}{2}}$$

$$m_0 = \frac{\text{K}}{t_d\sqrt{e}}$$

where K = Total effort in person years.

m = No. of persons at peak

t_d = Delivery time in years.

Now, consider once again,

$$m'(t) = \frac{d^2y}{dt^2} = 2Ka\, e^{-at^2}(1 - 2at^2)$$

Now, for $t = 0$,

$$m'(0) = 2Ka = \frac{2K}{2t_d^2} = \frac{K}{t_d^2}$$

or $$m'(0) = \frac{K}{t_d^2} = D = \text{Difficulty (person/year)}$$

but $$m_0 = \frac{K}{t_d\sqrt{e}} \text{ (just proved)}$$

$\therefore$ $$D = \frac{K}{t_d^2} = \frac{m_0\sqrt{e}}{t_d} \text{ proved.}$$

Example 5. *Prove the following relations:*

(*a*) *Mean,* $\mu = t_e = \dfrac{a + 4m + b}{6}$

(*b*) *Standard deviation,* $\sigma = \dfrac{b - a}{6}$

(*c*) *Variance,* $V = \left(\dfrac{b - a}{6}\right)^2$

where t_e, σ, *a, m, b and V have their usual meanings.*

Solution. The project managers considers *a*, *m*, *b* of each activity and uses the following relations for estimations.

(*a*) *Mean,* $\mu = t_e = \dfrac{a + 4m + b}{6}$

(*b*) Standard deviation, $\sigma = \dfrac{b - a}{6}$

(*c*) Variance, $V = \left(\dfrac{b - a}{2}\right)^2$

Let 'a' denote optimistic time
'b' denote pessimistic time
'm' denote mostly likely time
'σ' denote standard deviation
't_e' denote single time estimate
'V' denote variability of an activity.

Then, these can be marked in the graph given below:

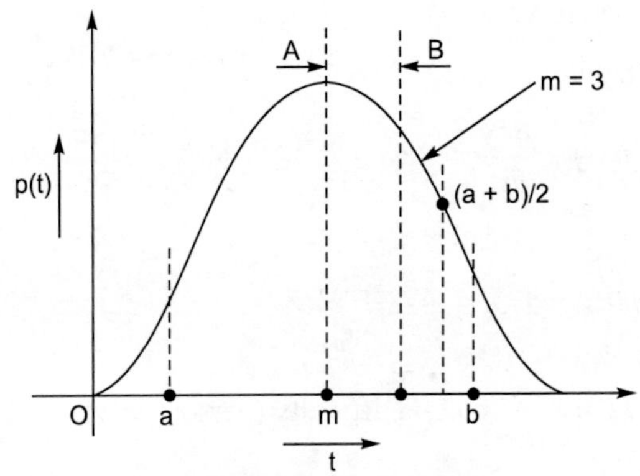

Fig. 11.1 β-curve for PERT.

Let 'A' represent mean. Also mark another point 'B' that is midway between *a* and *b*.

∴ From the statistical characteristics of β-curve we have

$$t_e = m + \frac{1}{3}AB \qquad ...(iv)$$

But

$$AB = \left(\frac{a+b}{2} - m\right) \text{ (from curve)} \qquad ...(v)$$

Substituting Eq. (*v*) in Eq. (*iv*), we get

$$t_e = m + \frac{1}{3}\left[\frac{a+b}{2} - m\right] = \frac{a + 4m + b}{6} \qquad \text{(where } a < m < b\text{)}$$

$$\therefore \quad t_e = \frac{a + 4m + b}{6}. \textbf{ Proved.}$$

Similarly, from curve given above, we find that

$$(b - a) = \pm 36$$

or $$6\sigma = (b - a) \quad \text{or} \quad \sigma = \frac{b - a}{6}.$$ **Proved.**

We also know that variance, $V = \left(\frac{b - a}{6}\right)^2$.

Now, as per the Central Limits Theorem, the standard deviation of the PERT network is the sum of standard deviations of the activities along the critical path.

∴ Variance, $$V = \sum_{i=1}^{n} \left(\frac{b - a}{6}\right)^2$$

and Standard deviation of PERT network (σ) is given by the formula

$$\sigma = \sqrt{\sum_{i=1}^{n} \left(\frac{b - a}{6}\right)^2}$$

where $i \leftarrow 1$ to n and are the activities, 1, 2, 3, ..., n which are lying along the critical path.

APPENDIX H

BIBLIOGRAPHY

1. Software Project Management by Mike Cottavell and Bob Hughes; Tata McGraw Hill, 2006.
2. Software Project Management by Royce; Pearson Education, Asia.
3. Software Project Management by Kelkar; PHI, 2007.
4. Software Engineering Project Management by Richard H. Thayer and Edward Yourdon; Wiley India, 2006.
5. Software Project Manual by Rishabh Anand; HCL Technologies Ltd., 2011.
6. Software Project Management by Manish Varshney and Ajeet Kumar; A.B. Publication, 2011.
7. Software Engineering – A Manager's Guide by Roger Pressman; TMH, 2006.
8. www.acm.org
9. All Project Management : www.allpm.com
10. Software Project Management by Rajiv Chopra; S.K. Kataria & Sons Ltd, 1st Edition 2011.
11. Software Project Management by Pankaj Jalote; Pearson Eduction Asia.
12. Software Project Management Guide by Rishabh Anand; HCL Technologies Ltd., 2012.
13. Software Engineering and Training Manual Guide by Rishabh Anand; HCL Technologies Ltd., 2012.
14. Prince 2 Foundation Manual by Rishabh Anand; HCL Technologies Ltd., 2012.

FREQUENTLY ASKED QUESTIONS

1. What do you understand by the term 'Software Project Management'? What are the control problems in producing the requirements specifications for a system?
2. Define the term 'Software Project? How is it different from other project? Describe the various activities involved in software project design and implementation.
3. (*a*) How the software project is different from other projects? Describe various phases of software project life cycle.

   (*b*) What is SRS? Describe the following aspects of requirements specification, traceability and unambiguous.
4. What is the difference between requirement definition and description (RDD) and SRS? What roles do they play in project and product development life cycle?
5. What are the activities involved in the planning of a software project? Discuss them.
6. Describe FP analysis.
7. Write short notes on the following:

   (*a*) RAD

   (*b*) Spiral model.
8. Explain Albrecht function point analysis for finding the effort required to complete a software project. What are the problems associated with over and under estimation?
9. Write short notes on following:

   (*a*) Gantt charts

   (*b*) Product breakdown structure

   (*c*) Precedence network.
10. How network charts help in scheduling a project? What is critical path?
11. Suggest a contingency action plan for the following risks:

    (*a*) Funds budget may fall short of 10% in the last phase of life cycle.

    (*b*) System designer may have to leave for USA after 2 months.

    (*c*) Software requires interface with other technologies on which the project team has no experience.
12. Differentiate between risk planning, risk avoidance, risk reduction and risk acceptance.
13. What do you understand by the term resource allocation? Discuss its nature, requirements and scheduling.
14. What do you understand by project evaluation and estimation? Discuss the different phases of estimation and various estimation methodologies
15. Describe EVA.
16. List and discuss the responsibilities of the software managers.
17. Define contract and contract management.
18. What do you understand by the term quality? Define software quality and discuss its importance. Is SQA an important activity?
19. Write short notes on:

    (*a*) ISO 9126

    (*b*) Product *vs* Process Quality Management.
20. Describe any software project management software which you have studied or worked on.

VIVA-VOCE QUESTIONS

1. What is a project?
2. What are the characteristics of a project?
3. What is the difference between software projects and other types of project?
4. Define Contract Management.
5. Define Panning.
6. What are the phase in software development life cycle?
7. Define Requirement Analysis.
8. What is meant by qualification testing?
9. What is the difference between Information Systems and Embedded System?
10. Differentiate objectives *vs*. products.
11. What is management?
12. Define project stake holders.
13. Define management control.
14. Define project evaluation.
15. How to review and publicize plan?
16. What is meant by programme?
17. Define vision statement.
18. What is blueprint?
19. Define development costs.
20. Define resource allocation.
21. What are the objectives of activity planning?
22. What is WBS?
23. Define Risk.
24. Define Risk Identification.
25. Define Risk Planning.
26. Define Risk Assessment.
27. What is resource planning?
28. What is monitoring?
29. What is control?
30. Define Gantt Chart.
31. Define Slip Chart.
32. Define Motivation.
33. Define Leadership.
34. Define quality.
35. What is quality cost?
36. What is external failure?
37. What is quality improvement?

38. What is "software safety"?
39. List at least seven deliverables in a software project.
40. What is ESTIMACS?
41. What is activity float or free float.
42. What is Brook's law?
43. What is risk leverage (RL)?
44. What is the role of the document editor?
45. Compare MS-PROJECT 2000 and MS-PROJECT 2002?

MODEL TEST PAPERS FOR PRACTICE

MODEL TEST PAPER-1

B.Tech. (CSE/IT) Seventh Semester

Software Project Management (CSE-403-F)

Time Allowed : 3 hours **Maximum Marks :** 100

Note: Attempt any five questions in total, at lease one question from each section. Question no. 1 is compulsory. All questions carry equal marks.

1. (*a*) Distinguish between normal, expected and exciting requirements.

(*b*) Define the terms – software, project and management.

(*c*) Suggest some approaches or methods to shorten the projects duration time.

(*d*) What are the difference between project management and line management?

(*e*) How is risk reduction different from risk mitigation?

(*f*) What is the purpose of configuration manager?

(*g*) Is SQA an important activity?

Section-A

1. (*a*) Comment on the statement:

(*i*) "Unambiguous requirement specification is an essential requirement for success of the project".

(*ii*) "Software projects may seem similar to other engineering projects but have some attributes which make them different".

2. Describe the contents list of project plan.

Section-B

3. (*a*) Explain Albrecht function point analysis for finding the effort required to complete a software project. What are the problems associated with over and under estimation?

(*b*) Describe FP analysis.

4. Describe the network planning model and the concept of critical path.

Section-C

5. (*a*) What is risk management and why it is important?

(*b*) You are given a project to develop a website for the college. Determine the following : Scope design, skill and development.

6. (*a*) Write a short note on EVA.

(*b*) Define contract and contract management.

Section-D

7. What do you understand by the term quality? Define software quality and discuss its importance.

8. (*a*) Write short notes on:

(*i*) ISO 9126

(*ii*) Product *vs* Process quality management.

MODEL TEST PAPER-2

B.Tech. (CSE/IT) Seventh Semester

Software Project Management (CSE-403-F)

Time Allowed : 3 hours **Maximum Marks :** 100

Note: Attempt any five questions in total, at lease one question from each section. Question no. 1 is compulsory. All questions carry equal marks.

1. (*a*) "Project Management is a conversion process". Comment.

(*b*) What is "software safety"?

(*c*) List at least seven deliverables in a software project.

(*d*) What is Brook's law?

(*e*) What is difference between version control and change control?

(*f*) What is ESTIMACS?

(*g*) What are the guidelines for facilitating communication?

Section-A

1. (*a*) Explain the terms—software, project and management in detail.

(*b*) Explain project as a system. Also explain project control and monitoring cycle.

2. Describe any software project management software which you have studied or worked on.

Section-B

3. What is the critical difference between the waterfall and spiral model?

4. Write short notes on:

(*i*) Software prototyping

(*ii*) Spiral model for software development.

Section-C

5. (*a*) Describe network chart. Explain with example, how they help in managing the schedule in a project.

(*b*) Discuss the common sources and types of risks in software development projects and strategies to deal with them.

6. (*a*) What do you understand by the term resource allocation? Discuss its nature, requirements and scheduling.

(*b*) List and discuss the responsibilities of the software managers.

Section-D

7. Explain why a high-quality software process should lead to high quality software projuct. Discuss possible problems with the system of quality management.

8. Define software quality. How ISO 9126 helps in achieving it?

MODEL TEST PAPER-3

B.Tech. (CSE/IT) Seventh Semester

Software Project Management (CSE-403-F)

Time Allowed : 3 hours **Maximum Marks :** 100

Note: Attempt any five questions. Question no. 1 is compulsory. Select one question from each section.

1. (*a*) What is backward and forward pass.

(*b*) "Project Management is a conversion process". Comment.

(*c*) Suggest some approaches or methods to shorten the project duration time.

(*d*) What is the difference between version control and change control?

Section-A

1. Define the term Software Project? How is it difference from other project? Describe the various activities involved in software project design and implementation.

2. (*a*) Describe the contents list of project plan.

(*b*) Discuss the following briefly:

(*i*) Management Control

(*ii*) Requirement Specification.

Section-B

3. Write short notes on the following:

(*i*) V-Process Model

(*ii*) Rapid Application Development.

4. (*a*) What is Risk Management? How is it identified analyzed, reduced and managed? Explain briefly.

(*b*) Describe the network planning model and the concept of critical path.

Section-C

5. (*a*) Explain resource scheduling. How the proper resource scheduling can lead to effective cost cutting?

(*b*) Define contract and contract management.

6. (*a*) Describe EVA?

(*b*) List and discuss the responsibilities of the software managers.

Section-D

7. Write a short note on:

(*i*) ISO 9126

(*ii*) Product vs Process Quality Management.

8. Discuss the following briefly:

(*i*) Project 2000 or any other SPM software

(*ii*) Techniques to increase Software Quality.

MODEL TEST PAPER-4

B.Tech. (CSE/IT) Seventh Semester

Software Project Management (CSE-403-F)

Time Allowed : 3 hours **Maximum Marks :** 100

Note: Attempt any five questions. Question no. 1 is compulsory. Select one question from each section.

1. (*a*) Is SQA an important activity?

(*b*) What are the differences between Project Management and Line Management?

(*c*) How is risk reduction different from risk mitigation?

(*d*) What is ESTIMACS?

Section-A

1. Comment on the statement:

(*i*) "Unambiguous requirement specification is an essential requirement for success of the project".

(*ii*) "Software Projects may seem similar to other engineering projects but have some attributes which make them different".

2. (*a*) What is SRS? Describe the following aspects of requirements specification. Traceability unambiguous.

(*b*) Write a short note on Software Project Planning discussing its all aspects.

Section-B

3. (*a*) Describe FP analysis.

(*b*) Write short notes on:

(*i*) Spiral Modal

(*ii*) RAD.

4. (*a*) How network charts help in scheduling a project? What is critical path?

(*b*) Write short notes on the following:

(*i*) Forward and backward pass

(*ii*) Gantt Charts.

Section-C

5. Discuss the common source and types of risks in software development project and strategies to deal with them.

6. (*a*) What do you understand by the term resource allocation? Discuss its nature, requirements and scheduling.

(*b*) Define contract and contract management.

Section-D

7. Define Software Quality. How ISO 9126 helps in achieving it.

8. Explain the main features of MS-PROJECT 2000. Where and how it is used.

EXAMINATION PAPERS

B.E. Seventh Semester (CSE/IT) Examination, May 2011
Software Project Management (CSE-403-E)

Time allowed : 3 hours **Maximum Marks :** 100

Note: Attempt any five questions. All questions carry equal marks.

1. (*a*) Comment on the statement:

(*i*) "Unambiguous requirement specification is an essential requirement for success of the project". (7)

(*ii*) Software projects may seem similar to other engineering projects but have some attributes which make them different." (7)

(*b*) Describe the content lists of project plan. (6)

2. (*a*) Define the term stakeholder. Who are the possible stakeholders in a software project and what can be their interests? (10)

(*b*) Write short notes on following:

(*i*) Gantt Charts

(*ii*) Production breakdown structure. (5 + 5 = 10)

3. (*a*) Define the following terms:

(*i*) Net present value (5)

(*ii*) Cash flow analysis. (5)

(*b*) Consider the project cash flow projection figures at the end of each year given in the table below. Calculate:

(*i*) Payback period of each project.

(*ii*) Find return on investment on each project.

Year	*Project 1*	*Project 2*	*Project 3*
0	– 100000	– 1,00000	– 100000
1	10000	200,000	30,000
2	10000	200,000	30,000
3	10000	200,000	30,000
4	200000	200,000	30,000
5	100000	300,000	30,000
Net Profit	50000	100,000	50,000

(10)

4. Writ short notes on following:

(*a*) Software prototyping

(*b*) Spiral model for software development. (10 + 10 = 20)

5. Explain Albrecht function point analysis for finding the effort required to complete a software project. What are the problems associated with over and under estimation? (20)

6. Write short notes on the following:

   (*i*) Network planning model

   (*ii*) Forward and backward pass

   (*iii*) Identification of critical path

   (*iv*) Precedence network (5 + 5 + 5 + 5 = 20)

7. (*a*) Differentiate between risk planning, risk avoidance, risk reduction and risk acceptance. (10)

   (*b*) Define Software quality. How ISO 9126 helps in achieving it. (10)

8. (*a*) Differentiate between product and process quality management. (8)

   (*b*) Describe some mechanisms other than money to motivate people. (7)

   (*c*) Define contract and contract management. (5)

EXAMINATION PAPERS

B.Tech. Seventh Semester (CSE/IT) Examination, December 2012

Software Project Management (CSE-403-F)

Time allowed : 3 hours **Maximum Marks :** 100

Before answering the questions, candidates should ensure that they have been supplied the correct and complete question paper. No complaint in this regard, will be entertained after examination.

Note: Attempt any five questions. Q. no. 1 is compulsory. Select one question form each Section.

1. (*a*) What is backward and forward pass.

(*b*) Explain the place of software quality in project planning.

(*c*) What are the different activities covered by SPM.

(*d*) What is earned value analysis?

Section-A

1. (*a*) What is Software Project? Explain various types of software project briefly.

(*b*) Discuss the following briefly:

(*i*) Management Control

(*ii*) Requirement Specification.

2. Write a short note on software project planning discussing its all aspects.

Section-B

3. Write short notes on following:

(*a*) V-Process model

(*b*) Rapid application Development.

4. (*a*) What is Risk Management? How is it identified, analyzed, reduced and managed? Explain briefly.

(*b*) Describe the network planning model and the concept of critical path.

Section-C

5. Explain the following briefly:

(*a*) Cost scheduling

(*b*) Resource scheduling

(*c*) Creating critical path.

6. Discuss different type of contract, its stages and typical terms of a contract. Also describe the contract management.

Section-D

7. Write short note on:

(*a*) ISO 9126

(*b*) Project 2000 or any other SPM software.

8. Discuss the following briefly:

(*a*) Product vs Process quality

(*b*) Techniques to increase Software Quality.

EXAMINATION PAPERS

B.Tech. Seventh Semester (CSE/IT) Examination, May 2013

Software Project Management (CSE-403-F)

Time allowed : 3 hours **Maximum Marks :** 100

Before answering the questions, candidates should ensure that they have been supplied the correct and complete question paper. No complaint in this regard, will be entertained after examination.

Note: Attempt any five questions. Q. no. 1 is compulsory. Select one question form each Section. All questions carry equal marks

1. (*a*) What is Abrecht function point analysis?
(*b*) What is the importance of software quality?
(*c*) What are the different type of contract? Explain.
(*d*) Explain different type of software project.

Section-A

1. Define the term software project. How is it different from other project? Describe the various activities involved in the software project management.

2. Describe the following briefly :
(*a*) Project Infrastructure Identification
(*b*) Project scope and objectives.
(*c*) Resource allocation

Section-B

3. Describe the following models of SDLC
(*a*) Spiral Model
(*b*) Prototyping Model

4. (*a*) Discuss the common sources and type of risks in software development projects and strategies to deal with them.
(*b*) What are the objectives of Activity Planning? Also explain sequencing and scheduling activities.

Section-C

5. (*a*) What do you understand by the term resource allocation? Discuss its nature, requirement and scheduling.
(*b*) Write a short note on Visualising Progress.

6. Explain the role and importance of the following in an organization :
(*a*) Working in groups
(*b*) Decision making
(*c*) Motivation
(*d*) Leadership

Section-D

7. (*a*) What do you understand by the term quality? Define the software quality and discuss the importance.
(*b*) Explain the following : Product Vs Process quality management.

8. Discuss in detail any Software Project Management software you have studied.

EXAMINATION PAPERS

B.Tech. Seventh Semester (CSE/IT) Examination, December 2013
Software Project Management (CSE-403-F)

Time allowed : 3 hours **Maximum Marks :** 100

Before answering the questions, candidates should ensure that they have been supplied the correct and complete question paper. No complaint in this regard, will be entertained after examination.

Note: Attempt any five questions. Q. no. 1 is compulsory. Select one question form each Section. All questions carry equal marks

1. (*a*) What is earned value analysis?
(*b*) What is Abrecht function point analysis?
(*c*) Is SQA an important activity?
(*d*) What are the difference between Project Management and Line Management?

Section-A

1. Comment on the statement :
(*i*) "Unambiguous requirement specification is an essential requirement for success of the project."
(*ii*) "Software Projects may seen similar to other engineering projects but have some attributes which make them different."

2. Describe the following item briefly :
(*i*) Project Infrastructure Identification
(*ii*) Project scope and objectives.

Section-B

3. Write a short notes on :
(*i*) Spiral Model
(*ii*) RAD

4. (*a*) Describe the network planning and the concept of critical path.
(*b*) What is Risk Management? How is it identified analyzed, reduced and managed? Explain briefly.

Section-C

5. (*a*) What do you understand by the term resource allocation? Discuss its nature, requirements and scheduling.
(*b*) Define contract and contract management.

6. Explain the role and importance of the following in an organization.
(*a*) Working in groups
(*b*) Decision making
(*c*) Motivation
(*d*) Leadership

Section-D

7. Write a short note on :
(*i*) ISO 9126
(*ii*) Product Vs Process Quality Management

8. Discuss in detail any software Project Management software you have studied.

Index

I

I

M

N

O

P

R

S

T

U

V

W

Win Prizes

ATTENTION READERS

Win Prizes

We request for your frank assessment regarding various aspects of the book as given below:

SOFTWARE PROJECT MANAGEMENT

By: Rishabh Anand

Please Fill up the Sheet in Neat and Clean Writing

(i) What topics of your syllabus, important for your examination point of view, are not covered in this book ?

(ii) What are the chapters wherein the treatment of the subject matter is not systematic or organised or updated ?

(iii) Have you come across misprints/mistakes/factual inaccuracies in the book? Please specify the chapters and the page numbers.

(iv) What other books on the same subject have you found/heard better than the present book? Please specify in terms of price and quality.

(v) Further suggestions and comments for the improvement of this book.

(PS: If need be please attach a separate sheet to write your views)

OTHER DETAILS

(i) Who recommended you this book ? Please tick (✓) in the box near the option relevant to you.

☐ Teacher ☐ Friends ☐ Bookseller

....

....

....

....

(ii) Name of the recommending teacher, his designation and address:

....

....

....

....

(iii) Name and address of the bookseller you purchased the book from :

....

....

....

....

(iv) Name and address of your institution (Please mention the University or Board as the case may be)

....

....

....

....

(v) Your name and complete postal address:

....

....

....

....

(vi) Course you are enrolled in:

....

....

....

....

The best assessment will be awarded each month: The award will be in the form of our publications as decided by the Editorial Board.

Please mail the filled up coupon at your earliest to:

S.K. KATARIA & SONS®
4885/109, Prakash Mahal, Dr. Subhash Bhargav Lane,
Opposite Delhi Medical Association, Daryaganj, New Delhi–110002 (INDIA)
Phone: +91-11-43551243; *Mobile:* +91-9871775858, 7678425209
e-mail: katariabook@gmail.com; katariabooks@yahoo.com
Website: www.skkatariaandsons.com